D1150418

The Genetic Basis of Development

TERTIARY LEVEL BIOLOGY

A series covering selected areas of biology at advanced undergraduate level. While designed specifically for course options at this level within Universities and Polytechnics, the series will be of great value to specialists and research workers in other fields who require a knowledge of the essentials of a subject.

TERTIARY LEVEL BIOLOGY

The Genetic Basis of Development

ALISTAIR D. STEWART, B.A., Ph.D.

Lecturer in Chemical Pathology
and Associate Lecturer in Genetics
University of Leeds

and

DAVID M. HUNT, B.Sc., Ph.D.

Lecturer in Genetics
Queen Mary College
University of London

Blackie

Glasgow and London

Blackie & Son Limited
Bishopbriggs
Glasgow G64 2NZ

Furnival House
14–18 High Holborn
London WC1V 6BX

British Library Cataloguing in Publication Data
Stewart, Alistair D.
 The genetic basis of development.—(Tertiary level biology)
 1. Developmental biology   2. Genetics
 I. Title II. Hunt, David M. III. Series
 574.3   QH491

ISBN 0–216–91161–3
ISBN 0–216–91160–5 Pbk

Filmset by Advanced Filmsetters (Glasgow) Limited
Printed in Great Britain by
Thomson Litho Ltd., East Kilbride, Scotland

Preface

This book discusses the significance of genetical concepts and techniques in developmental biology. It is aimed at students in their final undergraduate year, and at researchers at all levels who do not have genetical expertise and would like to consider its relevance to their work. We have assumed a previous understanding of eukaryotic genetics such as would be obtained from most broad introductory genetics courses. There is a strong emphasis on the description of genetic systems. Specific experiments are then evaluated in relation to the conceptual analysis of the developmental process which emerges as the book proceeds.

Most books are as notable for what they leave out as for what they include. In choosing the experiments, we have tried to keep a balance between conflicting requirements—for example, to consider the variety of different organisms, whilst ensuring that there is sufficient continuity from one chapter to the next for a picture of different developmental strategies to be constructed. In practice, this has meant that there is a stronger reliance on animals—Drosophila and the mouse in particular—than might otherwise be desirable. We feel that this is inevitable at the moment, given the amount of work done on these species. Chapter 4 (on the molecular biology of eukaryotic cells) is deliberately limited, as this subject is complex and is itself developing very rapidly. Here we have aimed merely to provide background information for other chapters, and to show that the use of gene cloning and sequencing techniques is likely to revolutionize our understanding of the organization of the genome and the regulation of gene expression. The final chapter considers one topic—sexual differentiation of mammals—in greater depth, and serves to illustrate the interrelationships between mechanisms acting at the different levels of organization which have been discussed separately in preceding chapters.

Several excellent books on developmental biology are now available, some of which use genetical examples. Although the regulation of gene expression is generally accepted as the basic paradigm for the understanding of development, many texts rely almost exclusively on the analysis of non-genetical studies to provide a bridge between the genome and the phenotype (the organism). This book gives expression to a growing awareness that a multi-disciplinary approach between geneticists, biochemists, developmental biologists and others is of crucial importance, and expounds the particular contribution of genetics in the context of the understanding provided by other disciplines. It is intended to complement the more traditional books on developmental biology, and can also be read in conjunction with books on eukaryotic molecular biology.

We would like to thank the friends, colleagues and students, too numerous to name, who have discussed these ideas with us. We are very grateful to Dr S. Baumberg for his helpful comments on a draft of chapter 4, and would like to thank Mrs Betty Sharp, Mrs Audrey Stewart, Mrs Doreen Jobbins and Mrs Carol Cusworth for typing. We are indebted to Mrs Joan Stratford for the production of the final manuscript.

In any book of this kind, there are bound to be some errors in description, and differences of opinion concerning the interpretation of experimental data. We would welcome any comments readers may wish to send to us.

A.S.

D.H.

Illustrations drawn by Terry Collins, Leeds Polytechnic

Contents

CHAPTER ONE

INTRODUCTION

This book is concerned with the processes of development in multicellular organisms. From before the turn of the century, clear descriptions have existed of the different stages in the development of many species of plants and animals, of their different life cycles, and of the role of cell movement in morphogenesis. In an effort to understand the mechanisms involved, many studies utilizing techniques for the experimental manipulation of embryogenesis have been carried out. The importance of the nucleus in development was clearly revealed in experiments such as those undertaken by Boveri and Hämmerling. Boveri enucleated sea urchin eggs of one species before artificially fertilizing them with sperm from a related species differing in features such as pigmentation and the form of the pluteus larva. For these characters, the artificial hybrids developed according to the nucleus rather than the egg, indicating that the nucleus rather than the cytoplasm directs morphogenesis. Hämmerling also examined the respective roles of the nucleus and cytoplasm in the unicellular alga, *Acetabularia*. He grafted the stem of one species on to the rhizome (which retained the nucleus) of a different species (figure 1.1). A new cap was formed, intermediate in appearance between those of the two parental types. If this was cut off, a second cap regenerated, whose shape resembled that of the species which donated the nucleus, rather than the one which provided the bulk of the cytoplasm (as the stem of the hybrid).

The concept that the roles of cytoplasm and nucleus should be seen as an interaction between the two components of the egg, rather than one or the other representing a decisive influence in isolation, emerges clearly from a classical experiment carried out on the egg of the dragonfly *Platycnemis* by Seidel in 1929. When he made a constriction with a hair loop at the posterior end of the egg before cleavage nuclei had time to interact with

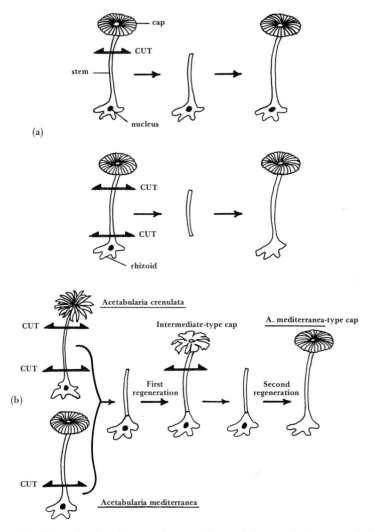

Figure 1.1   Hämmerling's grafting experiments with *Acetabularia* species: (a) removal of cap results in regeneration of a new cap whether a nucleus is present or not (b) rhizoid with a nucleus from *A. mediterranea* grafted to a stem of *A. crenulata* eventually produces an *A. mediterranea* type of cap.

this region, no embryo was produced from the anterior end. Alternatively, if the ligature was not pulled completely tight so that some cleavage nuclei entered this region, or if the constriction was made after the nuclei had

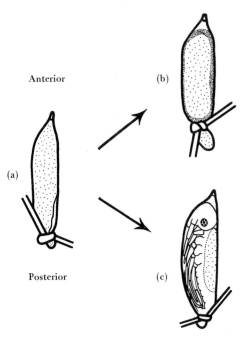

Anterior (b)

(a)

Posterior (c)

Figure 1.2 Seidel's experiments with the dragonfly, *Platycnemis*: (a) The posterior region of the egg is constricted; (b) With early constriction, the embryo fails to develop; (c) With late constriction, when any of the 128 blastoderm nuclei has passed into the posterior region, a complete embryo is formed.

entered and interacted with the posterior cytoplasm, development of the embryo proceeded normally (figure 1.2). From this it is apparent that a particular region of the peripheral cytoplasm at the posterior pole (the *activation centre*) is necessary to initiate embryogenesis, and that it can act only when a nucleus migrates into it. In fact, Seidel was able to show that any of the 128 nuclei present at the migratory phase of development were capable of mediating the action of this region.

Although these elegant experiments clearly demonstrated the importance of the nucleus in development, its mode of action remained almost completely unknown. Many experimental embryologists confined their attention to defining the role of cytoplasmic determinants and cell interactions in the control of developmental processes. However, following the rediscovery of Mendel's work at the beginning of the century, the science of genetics developed rapidly. Primary interest concerned the mode of gene action and the nature of the gene; but geneticists were also aware of the developmental anomalies produced by the mutations that they studied. The idea that development proceeded by the controlled switching on and off of particular gene loci in a sequence characteristic to each tissue was developed in the 1920's. Progress in the application of genetic techniques to the study of development owes much to the influence of R. Goldschmidt and C. H. Waddington. In his book *The Principles of Embryology* (1956), Waddington emphasized the genetic control of developmental pathways whilst giving a masterly summary of the findings of experimental embryology. Even so, these two aspects had largely to be presented separately. Among the reasons for this were the unknown nature of cytoplasmic determinants and the obscure mechanisms of tissue interactions on the one hand, and the inadequate models of the mode of gene action on the other. This dichotomy still persists. In many introductory texts, the regulation of gene expression is cited as the fundamental mechanism in development, but thereafter genetic studies are barely referred to. Nevertheless, genetic variation can be found affecting every phase of development. The aim of this book is to show how the manipulation of the genome has been combined with the experimental methods of biochemistry, endocrinology, physiology, immunology and classical embryology to provide new insights into the control of developmental processes.

The next two chapters examine the evidence for the constancy of the genome in development and the complementary studies on the production and role of cytoplasmic determinants in the egg. The following chapters look at the mechanisms of control of gene expression in eukaryotes. This is an area that has benefited considerably from the application of the new techniques of gene cloning and restriction enzyme analysis. The essential features of these techniques are reviewed and their application to the study of developmental regulation is examined. The organization of developmental processes in the embryo is then considered—whether developmental programmes are autonomous to a cell or group of cells in a tissue, or whether they depend on interactions with other cells or tissues. The interaction of cells is also discussed from a mechanistic standpoint. Finally, the book considers one complex developmental system where these processes can be seen in relation to each other.

CHAPTER TWO

THE CONSTANCY OF THE GENOME

Differential gene action during development could be achieved by two basic mechanisms. All cells might receive a complement of genetic information identical to that of the original fertilized egg. Different sets of gene products could then be formed if different regions of this constant genome are transcribed into RNA in different cells, or if there are differences in the processing of the transcribed RNA into mRNA, in its translation into protein, or in the activation (or degradation) of the protein. Alternatively, the various cells could receive different genetic complements, involving either the loss of particular gene loci or of whole chromosomes, or the formation of multiple copies of loci or chromosomes. This process would itself have to be controlled in the different cell lines, giving rise to a number of different clones of cells. Differential gene expression could also occur within the various clones. This chapter discusses the experimental evidence for the constancy of the genome.

Experimentally, various tests have been devised to distinguish between the mechanisms of differential gene expression and chromosomal differentiation. Broadly speaking, these tests either seek to examine the chromosome complements of differentiated cells by cytological or biochemical means, or to establish whether a differentiated cell or nucleus still retains the genetic ability to participate in the differentiation of another cell-type (a functional test of pluripotency) or even another whole organism (a functional test of totipotency).

Karyotypes
In the great majority of cases, mitosis during development results in differentiated cells with nuclei containing the same number of chromo-

somes (visible only during cell division) with identical morphology in the different tissues when viewed with the light microscope. Likewise, the DNA content per cell of different tissues is usually identical within experimental error. There are also exceptional cells which contain only half

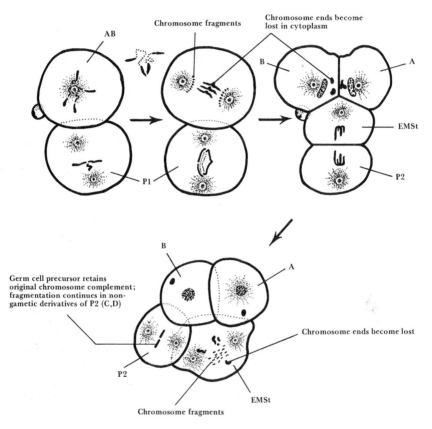

Figure 2.1   Cell lineage: dorsal cell family (AB) → most of ectoderm; EMSt → endoderm, anterior mesoderm, rudiment of stomadeum;

$$P_2 \rightarrow P_3 \rightarrow P_4 \rightarrow \text{germ cells}$$
$$\downarrow$$
$$D \rightarrow \text{posterior mesoderm}$$
$$C \rightarrow \text{posterior ectoderm}$$

Chromosome fragmentation in the early development of the nematode, *Ascaris* sp., as described by Boveri in 1899.

the number of chromosomes and DNA (such as spermatozoa), cells which lose their nucleus when differentiated (mammalian erythrocytes and plant xylem cells) and cells which are multinucleate or polyploid (mammalian myoblasts and liver, tobacco pith and plant roots). These changes, although a feature of the differentiation of those cells, cannot in themselves result in the formation of different gene products in different cells as they affect all loci equally.

In some cases, differential distribution of chromosomes is seen early in development. In nematode worms such as *Ascaris*, the cell line which gives rise to the germ cells and the gametes retains the full chromosome complement (by definition), whereas fragmentation and loss of satellite DNA sequences (see p. 13) from the ends of the chromosomes (Moritz and Roth, 1976) occurs in the other cell-lines in successive cell divisions (figure 2.1). The reduced chromosomal complement in the somatic tissues apparently remains stable during subsequent development. However, the functional consequences of the differences in karyotype between germ cells and somatic tissues are not understood, and the differences between the various somatic tissues cannot be attributed to the differential chromosome loss. In experiments where the egg of *Ascaris* was subjected to centrifugation in order to redistribute the cytoplasm, it was found that components of the cytoplasm localized at the vegetal pole protect the nuclei from chromosome loss and diminution. A similar phenomenon is seen in the gall midge *Wachtiella persicariae* (Geyer-Duszynska, 1959), and in the fungus gnat *Sciara coprophila* (Crouse, 1960) where certain entire chromosomes of the full complement are lost in somatic tissues only, and in the development of the *Sorghum* plant (Darlington and Thomas, 1941) where the supernumerary *B* chromosomes (which do not pair at meiosis) are lost from the root tissues, but are retained in some shoot tissues, and are transmitted to the next generation by the gametes formed in the flowers. However, these few cases of variation in the karyotype of different tissues are exceptions to the general rule of regular mitosis in plant and animal development.

Polytene chromosomes

In the Diptera, unusually large chromosomes, called polytene chromosomes, can be seen with the light microscope even during interphase. These occur in various tissues, such as the salivary gland, Malpighian tubule, rectum, mid-gut, oesophagus, tracheal wall and muscle of the larva, the bristle socket and footpad epidermis of the pupa, and Malpighian tubules

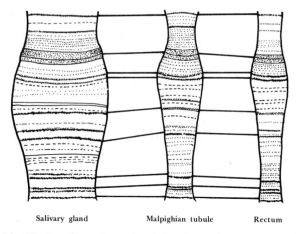

Salivary gland   Malpighian tubule   Rectum

Figure 2.2   Identification of homologous bands in polytene chromosome III of *Chironomus* (Redrawn from Fischberg and Blackler).

and ovarian nurse cells of the adult (figure 2.2). Polytene chromosomes arise by the process of endomitosis whereby chromosome replication proceeds in the absence of strand separation and cell division. The homologous chromosomes are initially paired and remain so during replication. Polyteny therefore provides a mechanism for the production of multiple gene copies in the cell, permitting a high rate of synthesis of particular gene products. Unpaired polytene chromosomes have been seen in some species of other insects (e.g. *Collembola*), in ciliates, and in the bean (*Phaseolus*), but the dipteran salivary gland chromosomes have been the most intensively studied. The individual chromatid strands of the chromosome are in a greatly extended state (compared to a normal mitotic chromosome) and are all precisely longitudinally aligned. When *Drosophila* chromosomes are stained with suitable dyes, about two thousand characteristic transverse bands or chromomeres can be seen with the light microscope. As this number corresponds approximately with the estimated number of structural loci coding for enzymes and other polypeptide chains in these insects, it has been suggested that each boundary between a band and an interband may contain a structural locus for a protein. More precisely, Judd and colleagues (1972) have estimated the number of genes in a small region of the X chromosome containing a total of 15 chromomeres. By isolating a large number of recessive lethal mutations mapping in this region, they were able to identify 16 comple-

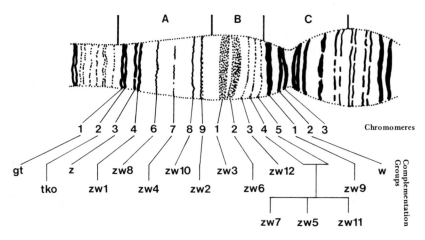

Figure 2.3 Assignment of complementation groups to chromomeres in the 3A–3C2 region of the X-chromosome of *Drosophila melanogaster*. The complementation groups are identified by the isolation of recessive lethals followed by complementation tests, and assigned to chromomeres by complementation tests with deletions which remove one or more bands from this region. (From Judd *et al.*, 1972).

mentation groups or genes. The analysis was subsequently extended to include non-lethal mutants (Judd and Young, 1973), and although additional genes were identified it is still probable that each chromomere contains either a single gene or at most a few loci (figure 2.3).

With normal karyotyping procedures, functionally significant chromosomal deletions or duplications could occur during development and yet escape detection. Although polytene chromosomes represent a nuclear specialization similar to polyploidy or multi-nucleation, they have been used to examine the possible occurrence of differential chromosomal changes during development with a much higher degree of resolution. Beerman (1952) carried out a comparison between different tissues of the banding patterns in chromosome 3 of the midge *Chironomus tentans* and found that he could identify the same bands in salivary glands, Malpighian tubules, rectum and mid-gut (figure 2.2). This and similar studies have been taken to provide strong evidence for the constancy of the genome.

Recently, though, Ribbert (1979) has reported that the banding patterns of polytene chromosomes from different tissues in the blowfly *Calliphora erythrocephala* bear little resemblance to each other, so that it may be necessary to revise this interpretation. The bands of this organism could be interpreted as reflections of functional activity, rather than as structural units. Alternatively, it is possible that the genome of this insect may not be constant.

In *Sciara* and related flies, spectrophotomeric studies have shown that extra DNA is produced at some bands, with a local geometric increase in DNA content (and presumably in gene copies) during late larval development (Crouse and Keyl, 1968). The nature and function of the extra DNA (referred to as DNA puffs by comparison with the RNA puffs discussed in chapter 4) is not clear, although some progress has been made in the identification of puff-dependent RNA and polypeptide synthesis (Winter and colleagues, 1977; Bonaldo and colleagues, 1979). DNA puffs are found in salivary gland cells but not in Malpighian tubules or gut (Breuer and Pavan, 1955) so that the genomes of these tissues do diverge during development. The significance of this process may be to allow the production of specific gene products required in large quantities in salivary glands.

Apart from the phenomenon of DNA puffing, the presence of all bands in a tissue does not necessarily mean that equal numbers of copies of all loci are present in the polytene chromosome. This question requires more precise and sophisticated techniques, such as molecular hybridization.

Molecular hybridization of the whole genome
Another means of examining the genetic complement of cells is provided by the technique of molecular hybridization. Two types of hybridization are possible: DNA/DNA and DNA/RNA. In each case, the principle behind the technique is the same (figure 2.4). The DNA double helix contains two complementary strands, and RNA molecules are complementary to the DNA strand from which they were transcribed. DNA can be sheared into short lengths and the complementary strands separated (melted) by heating in a salt solution at 80–90°C. If such separated strands are then incubated under suitable conditions, they will reanneal according to the specific base pairing that is possible. Non-complementary sequences will not hybridize. The extent of complementarity required for two strands to reanneal to form a stable hybrid is dependent on the experimental conditions (i.e. temperature, salt concentration). Hybridized (double-stranded) and unhybridized (single-stranded) nucleic acid can be separated by a variety of techniques that depend on their differential binding to an inert matrix. For example, if a solution of single- and double-stranded DNA is passed over a column of hydroxyapatite (a calcium phosphate complex) at low salt concentration, the double-stranded

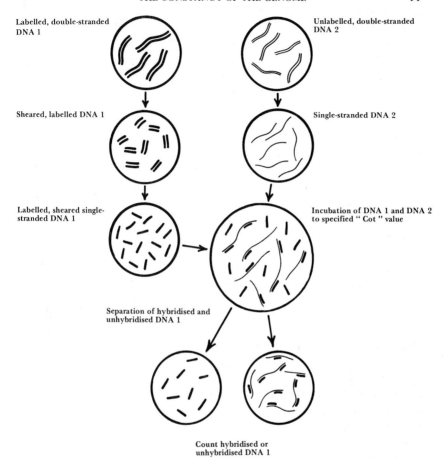

Labelled, double-stranded
DNA 1

Unlabelled, double-stranded
DNA 2

Sheared, labelled DNA 1

Single-stranded DNA 2

Labelled, sheared single-
stranded DNA 1

Incubation of DNA 1 and DNA 2
to specified " Cot " value

Separation of hybridised and
unhybridised DNA 1

Count hybridised or
unhybridised DNA 1

Figure 2.4 The principle of nucleic acid hybridization techniques. Labelled RNA or cDNA may be used instead of the labelled, double-stranded DNA 1 in this example. The hybridized and unhybridized DNA 1 may be separated by a variety of techniques.

molecules will be retained on the column whereas the single-stranded molecules will pass through. By increasing the salt concentration, the duplex DNA can be subsequently eluted from the column (figure 2.4). If one partner in the hybridization reaction is labelled isotopically, then the degree of reassociation can be readily measured.

The rate of hybridization will depend on the number of different sequences present and on the total concentration of nucleic acid. For comparability, the extent of hybridization is usually plotted against the

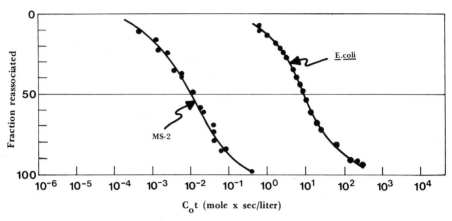

Figure 2.5 Hybridization of nucleic acids from MS-2 virus and *E. coli* (after Britten and Kohne).

product of concentration and time (the so-called Cot value). Cot values vary over several orders of magnitude and are therefore expressed on a logarithmic scale.

If we compare two organisms with very different genomic DNA contents, *E. coli* with 4.5×10^6 nucleotide pairs and the virus MS-2 with 4×10^3 nucleotide pairs (figure 2.5), we can see that the Cot value at which 50% reassociation will occur (called $Cot_{1/2}$) is correspondingly lower for MS-2 than for *E. coli*. The difference in $Cot_{1/2}$ is therefore a measure of the number of different DNA sequences present. The larger the genome, the greater the variety of sequences and the more difficult it will be for strands to find their complementary partners. If we consider the average mammalian genome to be about 1000 times larger than that of *E. coli*, we could expect mammalian DNA to reassociate with a $Cot_{1/2}$ of about 10^4. In reality, a portion of mammalian DNA reassociates much faster than this (figure 2.6). The only explanation for this surprising behaviour is that this portion consists of sequences repeated many times in the genome. These repeated sequences will therefore hybridize at a much lower Cot value.

The DNA of most higher organisms can be divided roughly into three classes on the basis of sequence frequency: *unique* (non-repetitive) sequences with a $Cot_{1/2}$ of 10^3 or more, *intermediate repetitive* sequences with a $Cot_{1/2}$ of 10^0-10^2, and *highly repetitive* sequences with a $Cot_{1/2}$ of $10^{-1}-10^{-4}$. This highly repetitive DNA frequently has a different GC/AT ratio from the average for the genome; it can therefore be physically

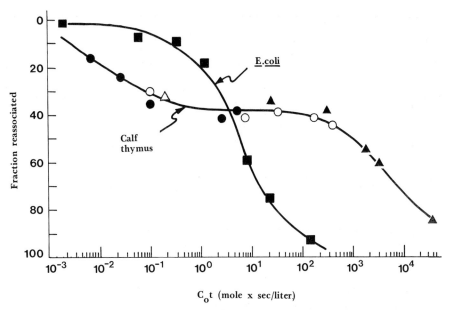

Figure 2.6   Reassociation of calf thymus DNA and *E. coli* DNA (after Britten and Kohne).

separated from the rest of the DNA by density gradient centrifugation (figure 2.7). GC-rich sequences with an increased buoyant density separate as a "heavy" satellite band, and AT-rich sequences with a reduced buoyant density separate as a "light" satellite band. Analysis of such bands shows that they consist of short sequences (5-378 bases long) in very large numbers of repeats (Goldring and colleagues, 1975), accounting for perhaps 10 % of the total genomic DNA.

Highly-repetitive DNA consists therefore of clusters of identical (or related) short sequences repeated more than 10^5 times. With the exception of the RNA associated with certain lampbrush chromosome loops of *Triturus* oocytes (Varley and colleagues, 1980) there is no evidence that these sequences are transcribed *in vivo*. It is nevertheless possible to prepare a highly radioactive RNA transcript by *in vitro* methods and this transcript can then be used as a probe to identify the chromosomal location of these sequences. Denatured DNA of chromosome preparations will hybridize this complementary RNA *in situ*. Since these sequences are arranged in clusters of many thousands of repeats, there will be sufficient hybridized RNA to allow the sites of specific binding to be revealed by

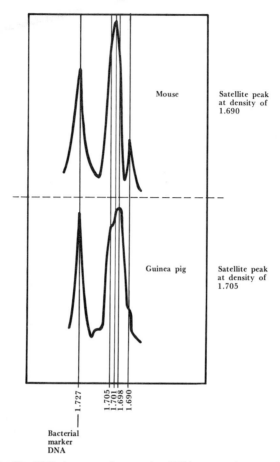

Figure 2.7   Satellite DNA in two rodent species. DNA preparations are loaded on to caesium chloride in an analytical centrifuge. Molecules will band in the caesium chloride according to their buoyant density. (After McConaughty and McCarthy).

autoradiography. Using this approach, Jones and Robertson (1970) found that these sequences are restricted to a few chromosomal locations and are particularly associated with the centromeres. Related species tend to have similar satellite sequences, although on occasion the sequences can be quite different, even in closely related species (Brutlag, 1980).

In contrast to the clustering of the highly-repetitive sequences, the intermediate sequences consisting of 10^2–10^5 repeats of between 700 and 1400 bases are interspersed throughout the genome (Davidson and

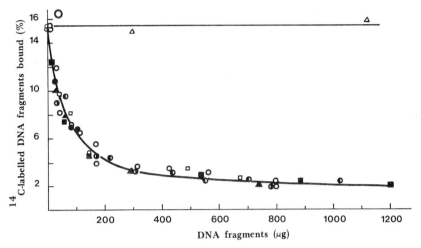

Figure 2.8 ^{14}C-labelled denatured DNA fragments from mouse L cells were incubated with denatured mouse embryo DNA immobilized in agar in the presence of increasing quantities of unlabelled DNA fragments from various mouse tissues and from *Bacillus subtilis*. DNA from mouse L cell (○), mouse embryo (●), brain (□), kidney (■), thymus (◑), spleen (◐), liver (▲), *B. subtilis* (△). Redrawn from McCarthy and Hoyer.

colleagues, 1973). The significance of these sequences remains obscure although a role in the regulation of gene action has been proposed. The unique fraction identifies sequences that are similarly distributed throughout the genome and are present only once (or perhaps a few times) per haploid set of chromosomes. As we shall see, gene sequences coding for polypeptides (i.e. structural genes) are to be found amongst this unique DNA fraction.

The DNA/DNA hybridization studies of McCarthy and Hoyer (1964) presented the first molecular test of the constancy of the genome. McCarthy and Hoyer compared the ability of DNA from various adult mouse tissues to compete with mouse L cell DNA for complementary binding sites on embryonic DNA (figure 2.8). In these experiments, the DNA from mouse L cells (a tumour cell line) is isotopically labelled, and the embryonic DNA is immobilized in agar. The identity of the different DNA preparations is shown by the ability of increasing quantities of unlabelled DNA (forming hybrids between differentiated cell DNA and embryonic DNA) to reduce the amount of labelled DNA (hybrids between L cell DNA and embryonic DNA) bound to the agar. Unlabelled DNA from mouse embryos also competes successfully (and to an equal extent in all cases) with the labelled L cell DNA. In other words, no differences between the hybridization behaviours of DNA from these different sources could be detected, whereas in control experiments, the hybridization behaviour of DNA from the bacterium *Bacillus subtilis* was quite distinct.

The identification of unique sequences during development

Since the unique sequences reassociate only very slowly, the above experiment does not establish whether any differences in these sequences may occur during development. For example, the loss of a particular structural gene would not be detected by this assay. This problem has been overcome by preparing the DNA for hybridization in a different way. If the mRNA coding for a particular protein can be isolated and purified, then it can be used as a template for the synthesis *in vitro* of complementary DNA (cDNA) from labelled precursors using a RNA-dependent DNA polymerase. (This enzyme, frequently called reverse transcriptase, carries out a similar function during the early stages of infection of RNA tumour virus such as avian myeloblastosis virus). cDNA with extremely high specific radioactivity can then be hybridized with DNA from a particular tissue, allowing the number of copies of the DNA sequence coding for that particular mRNA present in the genome to be estimated from the kinetics of the reaction. Thus, Sullivan and colleagues (1973) have concluded that there is one copy per haploid genome of the unique sequence coding for the egg-white protein ovalbumin both in hen oviduct cells (which produce large amounts of this protein) and in hen liver cells (which do not). To date, experiments of this type have shown that the DNA sequences coding for such diverse tissue-specific proteins as silk fibroin of the silk moth (Suzuki and colleagues, 1972) and the avian and mammalian globins (Harrison and colleagues, 1974; Packman and colleagues, 1972) are present in only one copy (or a very few copies) per haploid genome in all cells of the organism. The similarity of the embryonic and foetal globins to the adult types suggests, however, that these genes are closely related and have arisen by the duplication of an ancestral gene, followed by mutation. The globin genes are therefore an example of a multigene family of related sequences.

Multigene families

The existence of multiple copies of several genes has been known for many years. The 18S and 28S rRNA genes are arranged in tandem (figure 2.9) in

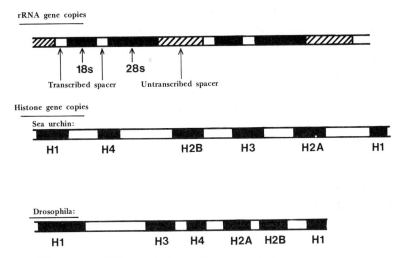

Figure 2.9 Maps of rRNA gene cluster and histone gene clusters.

a series of about 450–600 repeats per haploid genome (Brown and Weber, 1968), and the histone genes are similarly arranged as a unit coding for the 5 major histone proteins repeated from 100 to 1000 times depending on species (Kedes, 1979). The 5S RNA genes are also present in a cluster of many thousands of repeats (Brown and Sugimoto, 1973) and multiple copies of some of the tRNA genes are to be found either organized into clusters of different tRNA genes as in *Drosophila* (Yen and colleagues, 1977) or scattered throughout the genome as in yeast (Olson and colleagues, 1979).

Mammalian immunoglobulins are composed of two kinds of poly-peptide chain, the heavy (H) and light (L) chains. The L chains contain about 214 amino acids and consist of a variable region (responsible for antigenic specificity) and a constant region. H chains are rather larger (446 amino acid residues) but are otherwise similar in containing a variable region and a constant region, although in this case the constant region is about three times as long as in the L chain. The variable regions are coded by two quite separate genes, a family of V genes (which specify the major part of the variable region) and a family of related J genes (which specify the variable amino acid sequence of the polypeptide chain immediately adjacent to the constant region). In embryonic cells, the J genes are found as a cluster of sequences adjacent to the constant region (C) genes, whereas

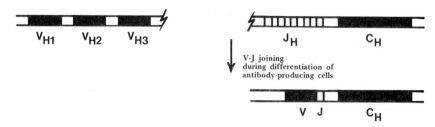

Figure 2.10   V, J and C genes for the heavy (H) immunoglobin chain, and the production of a functional, variable allele by joining of V and C genes during development.

the V genes, although linked, are at separate chromosome sites (figure 2.10). During differentiation of antibody-producing cells, V genes are translocated by intra-chromosomal recombination to a site near to the C gene. More specifically, a V gene becomes spliced to a J gene to form the DNA sequence coding for the entire variable region. Through transcription of the conjoined V, J and C genes then occurs to give a single primary transcript RNA for the entire H or L chain (Sakano and colleagues, 1979; Tonegawa and colleagues, 1978). In this way, therefore, a variety of different antibody molecules can be produced by a limited number of germ-line genes (additional variability is produced by somatic mutation in the V sequences). A functional V gene translocation occurs on only one of the two chromosomes and provides a molecular explanation for the phenomenon of allelic exclusion seen in antibody-producing cells, where individual cells produce only one type of immunoglobulin (i.e. from the association in the genome of only one V, J and C gene)—see pp. 123–32.

With the introduction of the techniques described in chapter 4 for the production of cDNA probes by gene cloning, many more multigene families have been identified. For example, the α- and β-tubulin genes in *Drosophila* are repeated four times per haploid genome. *In situ* hybridization of the cDNA to polytene chromosome preparations shows that each copy is at a different chromosomal site (Sánchez and colleagues, 1980). One such site corresponds to the map position of a male-sterility mutant gene affecting the production of a testis-specific β-tubulin (Kemphues and colleagues, 1980). The genes of this multigene family may therefore specify a number of different but related tissue-specific tubulins. Other examples of multigene families are the actin genes (Tobin and colleagues, 1980), and the insect chorion genes (Jones and Kafatos, 1980). (For a discussion of the evolutionary significance of multigene families, see Finnegan and colleagues, 1977).

The endochorion and exochorion of the insect egg contain more than 100 different proteins. These proteins are produced sequentially at high levels in the follicle cells. Prior to chorion gene expression in *Drosophila*, the follicle cells undergo several rounds of DNA replication in the absence of cell division. In addition, at least two of the chorion genes are further replicated or amplified to more than 10 times the haploid genome content (Spradling and Mahowald, 1980). In this way, the number of chorion gene copies is increased to meet the need for the rapid synthesis of chorion proteins. Only two other examples of gene amplification have yet been described; the amplification of the dihydrofolate reductase genes in drug-resistant mammalian cell lines (Dolnick and colleagues, 1979), and the amplification of the rRNA genes.

Ribosomal RNA gene sequences

Several examples of under- and over-replication of the rDNA sequences in specialized tissues have been reported.

In the polytene chromosomes of *Drosophila*, molecular hybridization reveals that only one of the two nucleolar organizers undergoes replication during endomitosis (Endow and Glover, 1979). This relative under-replication of the rDNA compared with non-polytene somatic nuclei was first observed by Spear and Gall (1973) (table 2.1) and, in view of the high amounts of polytene tissue in adult flies, almost certainly accounts for the apparent rectification or compensation of rDNA sequences reported by Tartof (1971) in flies carrying substantial deletions in one of the two nucleolar organizers. Under-replication is also seen in the heterochromatin (i.e. chromosomal material which remains condensed during interphase and is hence not transcribed) of polytene chromosomes which remains condensed and forms the chromocentre, a region where the hetero-chromatin of all four chromosomes becomes associated in a single mass. (The different classes of chromatin are discussed further in chapter 4).

Table 2.1 rDNA in diploid and polytene nuclei of *Drosophila melanogaster* (from Spear and Gall, 1973)

DNA	% rDNA ± SEM	Ratio of diploid to polytene
XX diploid	0.469 ± 0.016	6.0
XX polytene	0.078 ± 0.002	
XO diploid	0.264 ± 0.021	3.7
XO polytene	0.074 ± 0.007	

Amplification of the rRNA genes is also seen in some plants. In the cells of the onion, *Allium cepa*, hybridization experiments indicate massive amplification of the rDNA sequences (1125–3000 times the normal amount), with the extra DNA accumulating in and around the nucleolus (Avanzi and colleagues, 1973).

rDNA amplification has also been observed in the oocytes of many animal species including molluscs, crustaceans, insects, echinoderms, fish and amphibians (but not in mammals), and it has been especially well studied in the South African clawed toad, *Xenopus laevis* (Gall, 1969). In normal (wild type) animals a diploid somatic cell nucleus will contain two nucleoli, each with about 450 rDNA sequences. Nucleolar deficient mutants with reduced numbers of rDNA sequences are known. The *o-nu* mutant produces no true nucleolus and has less than 5% of the normal number of rDNA sequences (Brown and Gurdon, 1964). Not surprisingly, this mutant is lethal when homozygous. A second mutation, p^l-nu (Miller and Knowland, 1970) has a partial nucleolus and partial deletion of the rDNA (25% of the normal number of sequences). Since the oocyte is tetraploid (i.e. after chromosomal replication at prophase of the first meiotic division), it should normally contain about 1800 gene copies. In fact, the *Xenopus* oocyte contains 1000 or more nucleoli which are detached from the chromosome, and RNA/DNA hybridization shows that the rDNA sequences are correspondingly amplified. The multiple rDNA sequences can be visualized with the electron microscope (figure 2.11).

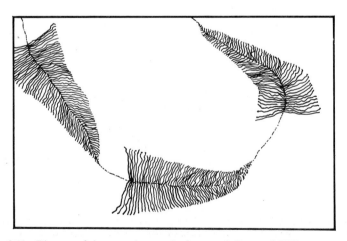

Figure 2.11   Diagram of electron micrograph of repeated ribosomal DNA sequences being transcribed in oocytes of *Xenopus laevis*. (Based on Miller).

Surprisingly, females heterozygous for the anucleolate mutant $(+/o)$, although containing half the normal number of nucleoli and rDNA genes in somatic cells, produce oocytes with the normal amplified number of nucleoli and rRNA genes (Perkowska and colleagues, 1968). This could be explained if (i) only one of the nucleolar organizers were involved in amplification, (ii) if some feed-back mechanism regulated the amount of rDNA produced, or (iii) if additional rDNA was kept as an episome in the

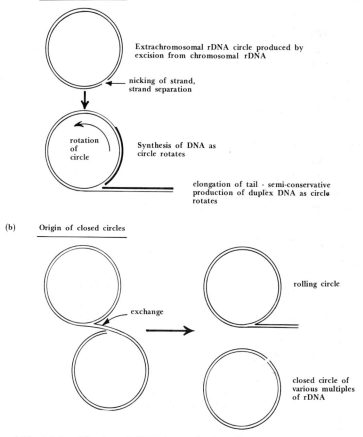

(a) Origin of rolling circles

Extrachromosomal rDNA circle produced by excision from chromosomal rDNA

nicking of strand, strand separation

rotation of circle

Synthesis of DNA as circle rotates

elongation of tail - semi-conservative production of duplex DNA as circle rotates

(b) Origin of closed circles

exchange

rolling circle

closed circle of various multiples of rDNA

Figure 2.12 (a) Amplification of rDNA by production of an extrachromosomal circle by chromosomal excision followed by rolling circle synthesis of extra copies of rDNA.
(b) Circularization of the elongating tail by recombination with the rolling circle will produce closed circles containing varying multiples of rDNA gene repeats.

germ cells only during embryogenesis of the mother to provide a template for rDNA amplification in her oocytes in the next generation. This last hypothesis would clearly give rise to maternal inheritance of the rDNA sequences and can be tested in reciprocal interspecies crosses between *Xenopus laevis* and *Xenopus borealis*. In such hybrids, the normal amount of rDNA is found in the nucleoli, but is only of the *Xenopus laevis* type (as determined by CsCl buoyant density and DNA/RNA hybridization) whichever species is the female parent (Brown and Blackler, 1972). This result would be expected on the first hypothesis, if the *Xenopus laevis* organizer is preferentially activated, and is also possible with the second hypothesis if the *Xenopus borealis* organizer is more sensitive to feed-back inhibition, but it is not consistent with an extra-chromosomal origin of the amplified rDNA sequences.

Electron microscopical examination of isolated amphibian oocyte rDNA reveals that a proportion is in the form of closed circles of discrete sizes corresponding to multiples of up to 4 or more tandem repeats of the 18S and 28S genes (Hourcade and colleagues, 1973). Rolling circles are also seen, suggesting a mechanism for rDNA amplification by elongation from an open 3′-OH terminus, displacement of a single-stranded tail, and semi-conservative synthesis to give duplex DNA (figure 2.12). In this way, additional rDNA can be produced continuously from an initial circle. The origin of this initial circle is less sure, although excision and circularization of chromosomal rDNA by a recombination-like process is possible.

During early cell division of the embryo, the extra nucleoli are lost, and rRNA synthesis depends on the normal nucleoli. It is found that the $+/o$ heterozygote with rather less than half the normal number of rDNA sequences can still make the normal amount of rRNA during early development, but genotypes with fewer DNA sequences cannot compensate and make abnormal amounts of rRNA (table 2.2). If it is

Table 2.2   Relative rDNA content and rRNA synthesis in various genotypes of *Xenopus laevis* (from Knowland and Miller, 1970)

	Nucleolar genotype				
	$+/+$	$+/p^l$	$+/o$	p^l/o	o/o
Relative amount of rDNA	100	60	45	23	6
Relative rate of rRNA synthesis	100	100	100	25–40	<2
Relative rate of rRNA synthesis per gene	100	166	222	112–175	<33

assumed from this that the rate of transcription in the single nucleolus of the $+/o$ heterozygote is the maximum possible, it can be calculated that it would take about 100 years to synthesize all the ribosomes found in the oocyte from four normal nucleoli without amplification to form multiple copies. The formation of many extra nucleoli in the oocyte is therefore thought to permit the production of the very large numbers of ribosomes found in this specialized cell.

The X- and Y-linked *bobbed* (*bb*) mutations in *Drosophila* (expressed in homozygotes as slow growth, short bristles and lethality) involve partial deletions of rDNA sequences of variable extent (similar to the p^l-nu mutation in *Xenopus*). In addition to the apparent increase in rDNA sequences in *bb* heterozygotes (Tartof, 1971), heritable increases associated with crossing-over at meiosis in the germ cells can also occur and this process is known as rDNA magnification (Ritossa, 1973). The accumulated rDNA from this process can then be reproduced and inherited indefinitely with the result that the phenotypic effects become less severe.

In summary, there is considerable evidence that rDNA sequences are over-replicated in a number of specialized tissues, and that amplification or replication from only one of the nuclelolar organizers may largely eliminate in heterozygotes the detrimental effects of genetic deficiencies in the rRNA genes.

Flax genotrophs—nuclear changes induced by environmental conditions
In particular varieties of flax (*Linum usitatissimum*) and tobacco (*Nicotiana rustica*), heritable changes can be induced during development by growing the plants under various environmental conditions. The best examples of this have been described by Durrant (1971) for flax. The original form of the plant (called Pl, for plastic genotroph) can be changed either to a form twice its weight (L: stable large genotroph) or to a form half its weight (S: stable small genotroph) if it is grown for five weeks after germination at a high temperature, with high nitrogen fertilizer or high phosphorus fertilizer as inducing agents respectively. The large and small genotrophs have remained stable for twelve generations in both environments, but they can both be induced to revert towards the plastic form by growing at a lower temperature during the critical five-week period. Complete reversion takes three generations, and the intermediate forms are also stable if grown at the higher temperature.

The genetic basis of these changes has been studied by making both reciprocal crosses and reciprocal grafts between the large and small forms.

There is no difference between the reciprocal crosses, and no changes are induced by grafting, so that there is no evidence for heritable changes in the cytoplasm, and the genetic difference between the forms appears to be nuclear (hence the term *genotroph*). This deduction is backed up by studies on nuclear DNA content as measured by Feulgen spectrophotometry. The stable increases or decreases in plant size during induction or reversion are correlated with stable increases or decreases respectively in chromosomal DNA of up to 16 % of the normal amount (Evans and colleagues, 1966), which can be seen to take place during the five-week induction period. Molecular hybridization has been used to characterize the differences in DNA content. The small genotroph has considerably fewer rDNA sequences (Cullis, 1977) than normal; as this difference is not nearly sufficient to account for the difference in total nuclear DNA content, other sequences must also be affected. The large genotroph seems to contain a new class of moderately-repetitive DNA sequences as well as a reduced proportion of highly-repetitive DNA sequences when compared with either the plastic or small genotrophs.

In flax, therefore, the changes that occur in DNA sequences during development are dependent on environmental conditions and apparently affect all tissues equally (and so are heritable). Consequently, they may be involved in the formation of differences between the genotrophs during growth and development, but not in differentiation of tissues within the plant. As the sensitivity to environmental induction is confined to particular varieties of flax, it will be interesting to learn more about the genetic factors controlling this responsiveness, as well as the mechanisms involved in the induction.

Pluripotentiality of differentiated cells

A large number of classical studies have attempted to demonstrate the pluripotentiality of cells from differentiated tissue. In animals, the equivalence of early cleavage stage nuclei is shown by Spemann's experiments on *Triturus* eggs. These showed that when a recently fertilized egg was constricted with a fine hair so as to leave only a narrow cytoplasmic bridge between the two halves, cleavage would proceed only in the half containing the nucleus. If at the 8- or 16-cell stage, one of the daughter nuclei migrated through the cytoplasmic bridge into the undivided blastomere, this then underwent cleavage. If the ligature was now pulled tight so that the two halves were separated, each half went on to develop into a normal dwarf embryo (figure 2.13).

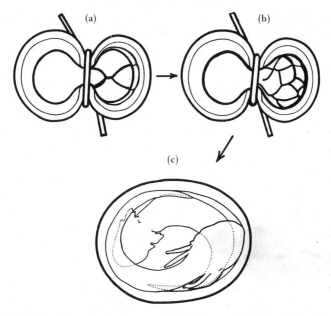

Figure 2.13   Totipotency of early cleavage nuclei in the newt. Shortly after fertilization, an egg is constricted with a loop of hair. (a) Cleavage occurs only in the half containing the nucleus; (b) the passage of a nucleus into the other half intiates cleavage; (c) two complete embryos may result showing that this early cleavage nucleus was still totipotent. (After Spemann).

Pluripotent tumours or teratocarcinomas arise spontaneously in mammals at low frequency from testicular germ cells before gametic differentiation has commenced (although the frequency in inbred strain 129 of the mouse may be as high as 32% of all male mice) or from parthenogenetic development of oocytes in the ovary, and are also produced at high frequency in the mouse following implantation of early embryo cells into ectopic sites (such as under the kidney capsule). Such tumours consist of many disorganized but differentiated tissues. The tissues are initially derived from undifferentiated malignant stem cells or embryonal carcinoma cells that are similar to the cells of the early embryo (Martin, 1980). The equivalence of embryonal carcinoma cells to embryonic cells is further demonstrated by their ability, when injected into the blastocoel of a normal blastocyst-stage embryo, to colonize the inner cell mass and to contribute to the tissues of the resulting adult mouse (Brinster, 1974; Mintz and Illmensee, 1975). More recently, Papaioannou and co-workers (1978) have shown that embryonal carcinoma cells cultured *in vitro* can also support normal development when placed into a normal embryonic environment.

In later development, the larval imaginal discs of *Drosophila* (from which many of the adult structures will develop during the pupal period) are still capable of a change in determined state (transdetermination) if additional growth is induced by culture of the disc in an adult host. This phenomenon is further discussed in chapter 7. Evidence for the pluri-potency of adult, differentiated animal cells is found in regeneration. The regeneration of an amputated limb of a urodele amphibian is preceded by the formation of a blastema from stump tissues. These blastema cells originate from differentiated tissues and some degree of dedifferentiation is therefore apparent. The developmental potential is however very much dependent on the origin of the blastema cell. Some, such as those from cartilage, retain a strong bias in regeneration towards the tissue of origin, while others (especially those with a connective tissue origin) are more labile and differentiate into a limited range of cell types of mesodermal origin, such as fibroblasts, muscle and cartilage (Namenwirth, 1974). Unequivocal dedifferentiation of specialized cells is seen in amphibia in the regeneration of the lens of the eye. In urodele amphibia, the surgical removal of the lens results in the conversion of iris cells to lens cells (metaplasia) and the development of a new lens (Yamada, 1977). Anuran amphibia such as *Xenopus laevis* can also regenerate a lens although in this case the regenerative capacity is restricted to the tadpole stage and the new lens cells are derived from the corneal epithelium (Freeman, 1963).

In several plant species including the carrot (*Daucus carota*), tobacco plant (*Nicotiana*) and the water parsnip (*Sium suave*), whole fertile plants have been obtained by culturing single cells from a variety of differentiated tissues under suitable conditions (see review by Vasil and colleagues, 1979). The evidence of totipotency from plant studies is therefore impres-sive. In contrast, an isolated animal cell from a differentiated tissue has never been successfully cultured to form a complete fertile animal. Only the more limited cases of pluripotency described above have been seen. This does not necessarily mean however that the nuclei of differentiated animal cells have undergone irreversible change. Rather, it indicates that the environment of the animal egg cytoplasm may be essential for the expression of this totipotence.

The totipotence of a nucleus can be tested in principle by removing it from its original differentiated cell and transplanting it into the appro-priate cytoplasmic environment, that is, an unfertilized egg whose own nucleus has been removed or inactivated. An anucleate egg does not develop or differentiate on its own, but a totipotent nucleus should be able to replace the original one so that a normal adult develops. Nuclear transplantation experiments of this kind were first carried out successfully in amphibians.

A number of problems can however arise in interpreting the results of this sort of experiment. A normal adult might result because the original nucleus was not successfully removed or inactivated. Conversely, abnormal individuals might develop if the transplanted nucleus were

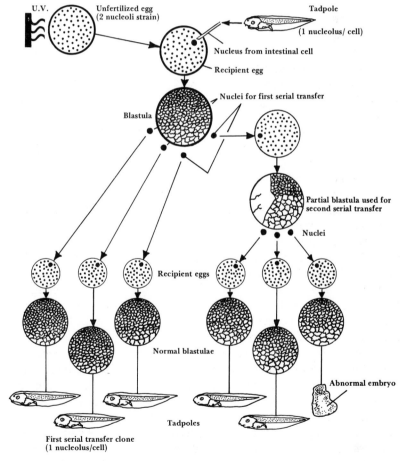

Figure 2.14 The serial nuclear transfer technique in *Xenopus*. (Modified from Gurdon).

damaged, if it did not respond adequately to its new environment, or if it were not totipotent. Gurdon and his colleagues, together with other groups, have largely overcome these difficulties in *Xenopus* by the use of genetic techniques and ingenious experimentation. Gurdon and Laskey (1970) used animals heterozygous for the nucleolar mutant described earlier in this chapter ($+/o$-nu) as donors, so they could be quite sure that the original nucleus with its two nucleoli had been inactivated. They also used serial transfers, which in effect clones the original nucleus and thus permits an assessment of its potency even if a proportion of the nuclei are damaged during transfer (figure 2.14).

Nuclear transplantation from blastula or gastrula endoderm cells is strikingly successful; in *Xenopus* more than half the eggs develop into normal tadpoles and some develop into normally sexually-mature adults. Fertile adults have also been obtained with donor nuclei from the rapidly dividing tadpole intestinal epithelial cells (Gurdon, 1968; figure 2.15). Nuclei from adult differentiated cells can also support considerable development but only if explants of the adult tissues are made and donor

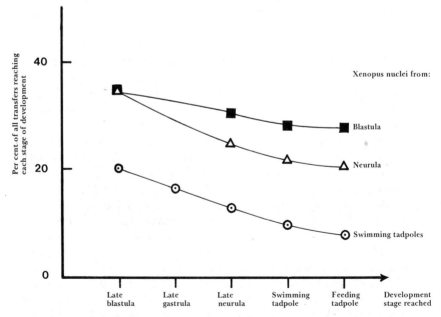

Figure 2.15 The survival of embryos receiving nuclei from different stages of development in *Xenopus*. (Redrawn from Gurdon).

nuclei taken from the cultured cells that grow out from it (Gurdon and Laskey, 1970). Although the cells that grow out from most explants are not normally recognized as differentiated cells and may be of a "fibroblast" type, this is not the case for adult skin explants where there is considerable evidence that the cultured cells produced are committed skin cells (Gurdon, 1974).

The serial transfer of nuclei derived from a "transplant" embryo indicates that when abnormal development occurs in an embryo, other embryos derived from the same nuclear clone tend to show the same abnormality (i.e. cease development at the same stage). However, virtually all developmental abnormalities are associated with chromosome breakage which has been found to occur soon after nuclear transplantation (Gallien et al., 1963; Hennen, 1963). The restricted developmental potential of many donor nuclei probably arises therefore from their inability to complete DNA replication within the 1–2 hours that will elapse after egg activation by micropipette penetration and the first cleavage division. Most somatic cells divide very infrequently and take about 7 hours to replicate their chromosomes. It is not surprising, therefore, that the most successful donor nuclei are from embryonic cells that normally undergo rapid division. Chromosome damage may also account for the lack of fertility in many of the animals that do reach maturity. These chromosomal abnormalities do not however vary significantly in embryos with donor nuclei from different cell types, and are therefore not related to the process of differentiation.

Despite the interpretative difficulties, it is reasonable to conclude that differentiated cells in the frog contain pluripotent, or even totipotent nuclei. Nuclear transplantation has also been extensively investigated in *Drosophila* (see Illmensee, 1976). Preblastoderm nuclei from the anterior end of genetically marked embryos, when injected into the posterior pole of fertilized recipients, give rise to fertile chimaeric adults that may even produce progeny of the donor genotype (Okada and colleagues, 1974), indicating that under these conditions the preblastoderm nuclei are totipotent. In mammals, recent attempts to transplant nuclei have been dramatically successful. By the relatively simple but ingenious innovations of including cytochalasin B in the culture medium to "relax" the cytoskeleton of the egg and withdrawing the male and female pronuclei from the fertilized egg *after* the introduction of the donor nucleus with the same micropipette, thus avoiding two penetrations of the egg, Illmensee and Hoppe (1981) have successfully transplanted mouse embryonic nuclei into fertilized eggs. From 142 C57BL/6 eggs successfully injected with either

CBA/H-T6 or LT/Sv nuclei from inner cell mass cells, 48 developed in culture to the morula and blastocyst stages. From these, 16 were selected for transference to albino foster mothers (ICR/Swiss), together with 44 non-injected BALB/c or ICR/Swiss blastocysts. All the females became pregnant and gave birth to 32 albino and three pigmented young. By karyotypic and enzyme analysis, all three pigmented mice were shown to be identical to the nuclear donor strains, and two of the mice subsequently produced several normal progeny of the nuclear-donor phenotype. These experiments demonstrate unequivocally that nuclear transplantation can be successfully carried out in the mouse and that nuclei from the inner cell mass are totipotent.

Conclusion

The different types of evidence reviewed in this chapter must be considered in relation to each other. In searching for basic mechanisms, much reliance has to be placed on intensively studied systems and species. There is no consistent evidence *against* the view that differentiated plant cells and nuclei from differentiated animal cells are normally totipotent, and the evidence from molecular hybridization is also consistent with the idea that, with certain exceptions, all cells contain one, or a few, copies per haploid genome of each structural locus coding for a polypeptide chain. The totipotent nuclei may also contain DNA sequences repeated to a greater or lesser extent in different tissues and at different times. Polyploidy is one mechanism for increasing the number of gene copies but this is of course not selective for any particular sequence. The selective amplification of particular gene loci is seen in a few cases and has been particularly well documented for the rRNA genes in amphibian oocytes. During development, the chromosomes may themselves undergo some degree of differentiation (most obviously seen in polytene chromosomes and in the rearrangement of immunoglobulin V genes) which may be reversible to a variable extent in the novel circumstances following experimental manipulation, whereas the few known cases where chromosome loss occurs during normal development may represent an extreme example of irreversible chromosomal differentiation.

In early development, cytoplasmic signals from the egg are clearly of primary importance. For example, the amphibian nuclear transplant experiments show that the cytoplasm of the egg is effectively able to re-programme the donor nucleus and override any chromosomal change that may have occurred, while chromosome loss, as seen in *Ascaris* and

Wachtiella embryos, depends on components of the egg cytoplasm. Similarly, the production of whole plants from single differentiated cells depends on specific culture conditions acting through the cytoplasm of the cell. The development of different cell types from a single fertilized egg must depend on a series of precise nuclear-cytoplasmic interactions, leading to a reduction in developmental potential and culminating in cell differentiation. The nature of these interactions is the topic of the next chapter.

CHAPTER THREE

CYTOPLASMIC REGULATION
OF GENE EXPRESSION
DURING DEVELOPMENT

If the genetic complement of the nucleus remains essentially unaltered during early development, then differences in gene expression between cell types must originate in some cytoplasmic heterogeneity in the egg. Some evidence for the role of cytoplasmic factors has already been seen in the centrifugation experiments on *Ascaris* eggs in relation to chromosome loss during development and in the identification of the activation centre in *Platycnemis* eggs. However, the precise nature of this cytoplasmic information varies considerably amongst the different taxonomic groups.

In some algae (such as the brown seaweed, *Fucus*), the egg is initially unpolarized. Several hours after fertilization however, rhizoid development begins, and by the end of the first cleavage division, the two resulting cells are quite distinct. The site of rhizoid development is initially determined by the site of sperm entry but may be subsequently modified by a number of environmental factors such as pH, temperature, voltage, light and ionic (Ca^{2+} and K^+) concentration (Quatrano, 1978). In the mouse *Mus musculus*, all the cells in early cleavage are identical and apparently totipotent; their developmental fate seems to be determined by their position within the embryo. Early cleavage divisions in mammals are asynchronous and give rise to a cluster of cells called a morula. The developmental capacity of such cells has been extensively examined in embryo-reconstitution experiments. Kelly (1975) dissociated four-cell stage embryos and allowed each isolated cell to divide *in vitro* before surrounding it with cells derived from an 8-cell stage embryo. Such composite embryos or chimaeras show considerable developmental regulation and will complete their development if transferred back into the uterus of a prepared (pseudopregnant) foster-mother. By combining cells from embryos carrying different coat colour genes and electrophoretic

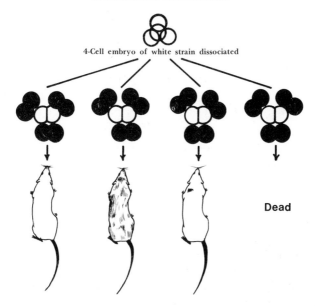

4-Cell embryo of white strain dissociated

Dead

Figure 3.1 Developmental potential of blastomeres dissociated at the 8-cell stage of the mouse. Each blastomere was allowed to divide and was then surrounded by blastomeres derived from a different strain. (Modified from Kelly).

enzyme variants, Kelly was able to show that the isolated cells are capable of contributing to both embryonic and extra-embryonic structures (figure 3.1). Genetic analysis of the resulting adult chimaeras showed that the isolated cells had also contributed to the germ-line. Similar results have been obtained by Gardner (1975) working with later blastocyst-stage embryos. Isolated inner cell mass cells, if introduced into the blastocoel of a second blastocyst, will colonize the host inner cell mass and contribute to a variety of tissues of the resulting chimaeric mouse. There is therefore no evidence for the existence of morphogenetic determinants in the mouse egg. Rather, it is the final position of the cell in the developing embryo that determines its developmental fate.

In land plants and in many animals, there is considerable evidence that cytoplasmic heterogeneity of the egg does affect cell determination. One such example comes from the pioneer work of Conklin on the ascidian *Styela partita*, and similar conclusions have been reached more recently by Whittaker and colleagues (1977) working with a related ascidian, *Ciona intestinalis*. In the egg of *Styela*, distinctive cytoplasmic regions can be identified. In the unfertilized egg, a central grey cytoplasm is surrounded by a cortical layer containing yellow lipid droplets and mitochondria, and a region of clear cytoplasm is to be found near the animal pole. At fertilization, these distinctive regions are rearranged so that the yellow and clear cytoplasms collect around the site of fertilization with the grey yolk granules now occupying the animal pole. During cleavage, these cytoplasmic regions are segregated into specific blastomeres which have been identified by cell lineage studies to be the precursors of definitive regions of the larva. The association of specific cytoplasmic regions is not in itself sufficient evidence that these regions carry morphogenetic determinants which play a causative role in development. Conklin's blastomere destruction experiments, however, provide compelling evidence that this is indeed the case. Embryos damaged in this way produce only incomplete larvae; the missing structures always correspond to the damaged blastomeres. Similar conclusions arise from experiments where the positions of these distinctive cytoplasms have been disrupted by centrifugation of embryos. In this case, development proceeds according to the new location of the cytoplasm and results in highly abnormal larvae.

A further example of the cytoplasmic segregation of morphogenetic determinants is found in the mollusc *Ilyanassa* (a marine snail). In this species, a lobe of vegetal cytoplasm (the polar lobe) forms at the first and second cleavage divisions. This lobe does not contain a nucleus, yet if mechanically removed, the resultant embryo always has defective mesoderm and mesodermal derivatives (figure 3.2). Detailed cell deletion experiments of Clement (1962) have shown that the cytoplasmic determinants originally contained within the polar lobe are progressively segregated out at subsequent cleavage divisions.

The surgical removal of a region of the vegetal cytoplasm of anuran amphibian eggs also results in developmental deficiency, in this case the loss of primordial germ cells with consequent sterility in the resulting adult

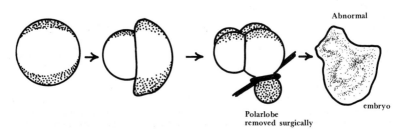

Polarlobe
removed surgically

Abnormal

embryo

Figure 3.2   Removal of the polar lobe at the 4-cell stage in *Ilyanassa*.

frogs. This region is therefore called the germ plasm. The grey crescent region of the amphibian egg has even more striking properties. Grafting this area to the prospective belly region of a second fertilized egg results in the appearance of a second, normal embryo alongside the normal embryo of the host egg, whereas removal of the grey crescent from an egg (or isolated part of an egg) results in failure to form an embryonic axis (figure 3.3). There is little doubt therefore that the different cytoplasmic regions

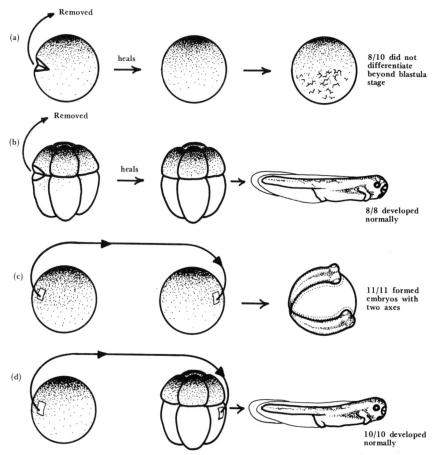

Figure 3.3 Cortical grafting of the grey crescent region of the egg in *Xenopus laevis*. (a) Effect of removal of grey crescent region at one-cell stage; (b) removal of grey crescent region at eight-cell stage has no effect; (c) grafting from one-cell stage to a second egg produces two embryonic axes; (d) grafting from the one-cell stage to a second egg at the eight-cell stage has no effect. (Modified from Curtis and Graham).

identified in these and similar experiments contain cytoplasmic components which affect differentiation and morphogenesis, presumably through their effect on gene expression.

Recently, there has been renewed interest in the nature of such cytoplasmic determinants. They might represent specific localized "instructions" with a purely local effect on nuclei in the relevant cells which sets an autonomous cell-specific developmental programme into action (the mosaic theory). Alternatively, they may act as points within the developing embryo (reference points or sites of "instruction") which affect other cells in relation to their distance from the site of the "instruction".

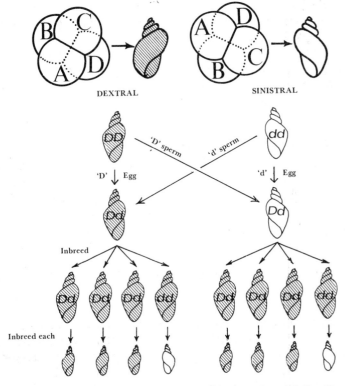

Fig. 3.4 (top) Relation between direction of cleavage and direction of shell coiling in the adult snail. (Redrawn from Waddington).

(bottom) Inheritance of the cleavage pattern and direction of shell coiling through four generations in the snail, *Limnaea*. (sinistral = left handed: dextral = right handed). The phenotype is determined by the maternal genotype. (Redrawn from Sinnott, Dunn and Dobzhansky).

(This distinction does not completely correspond to the distinction made in classical embryology between mosaic and regulative development, since embryonic regulation of differentiation according to the position of a new reference point may not always be possible if cells have already responded to the original reference point. When an experiment is performed some time after fertilization, a mosaic type of development would be apparent). As an egg initially without instructions (such as *Mus*) or with highly labile instructions (such as *Fucus*) develops, cytoplasmic determinants will become defined, either in the form of reference points or a mosaic. The possibilities are, of course, not exclusive; one egg could contain both types of determinant. Furthermore, gene expression in different cells at different distances from a reference point might respond to the instruction in qualitatively different ways, hence amplifying the initial positional differences into mosaic form during development.

The properties of an egg are determined during oogenesis. As the egg develops in the ovary, it is therefore subject to the action of maternal gene loci. We shall now consider some of the effects of maternal gene action in relation to these theories of developmental control.

Direction of shell coiling in the snail

A classical example of an effect of the maternal genotype on offspring, regardless of their own genotype, was described in the water snail, *Limnea peregra*, by Boycott and colleagues (1930). The shell and internal organs may be arranged in either a right-handed (dextral) or left-handed (sinistral) sense, and this character can be traced back to the cleavage pattern of the egg (figure 3.4). It was found that F1 individuals always resembled the female parent. Whatever their phenotype, these F1 individuals gave rise to uniformly dextral F2 individuals on inbreeding, whereas the F3 individuals, produced by further inbreeding, segregated 3 dextral : 1 sinistral (figure 3.4).

The puzzling observation here is that genetic segregation only occurs in the F3 and not in the F2. The direction of the coiling must depend therefore not on the individual's own genotype, but on the genotype (and *not* the phenotype) of its female parent. Developmentally, this is reasonable as the maternal genotype can influence the structure of the developing oocyte in the ovary before fertilization. The 3 : 1 ratio indicates that only one locus need be postulated, and that the dextral allele is dominant to the sinistral. This interpretation, first put forward by Sturtevant, is consistent with the data outlined above.

This phenomenon provides clear evidence that the arrangement of cytoplasmic determinants in the egg is genetically defined. They could be of either the mosaic or reference point type. The biochemical mechanism whereby handedness is defined by the genotype and expressed in the structure of the oocyte, surrounded by follicular nurse cells of the same genotype, is not understood.

The germ plasm in insects

During early embryogenesis in *Drosophila* and many other insects, cleavage is distinctive. The initial mitotic divisions are not accompanied by cytokinesis. After several such divisions, the nuclei migrate to the surface of the embryo to form the blastoderm. Cell membranes are immediately formed around the few nuclei which come to occupy the posterior pole plasm (see also p. 141). These pole cells are further distinguished by the presence of polar granules. The remaining nuclei undergo several further rounds of division before cell membranes appear. Different regions of the cytoplasm can be manipulated (surgically, or by irradiation with visible or ultraviolet light) before the nuclei have arrived to produce developmental defects that mimic or phenocopy the effects of maternally-expressed gene loci.

Illmensee and Mahowald (1974) have carried out transplantation experiments on the germ plasm of *Drosophila melanogaster* (similar to those on amphibians described earlier in this chapter) to determine if the cytoplasmic determinants for primordial germ cells in this posterior region of the egg are of the mosaic (autonomous) type. Using genetically marked flies to identify the source of the nuclei giving rise to functional gametes (figure 3.5), posterior pole cytoplasm was transferred to the anterior part of another egg. Nuclei of this egg moved into the transplanted cytoplasm and formed histologically normal primordial germ cells, so that the cytoplasm apparently contained autonomous determinants. If these cells were transplanted to the posterior region of a third egg, they formed viable gametes and transmitted their genetic markers to about 4% of the progeny in a test cross. This role of posterior germ plasm in germ cell induction is consistent with the experiments of Okada and colleagues (1974) who have shown that the germ plasm can be inactivated by UV irradiation: fertility is restored by injections of posterior germ plasm (containing polar granules) but not by anterior cytoplasm. The polar granules are rich in RNA; the inactivation of germ plasm by UV irradiation may depend on the degradation of this RNA. Recently, a basic protein of 90 000 daltons has been

GENOTYPE

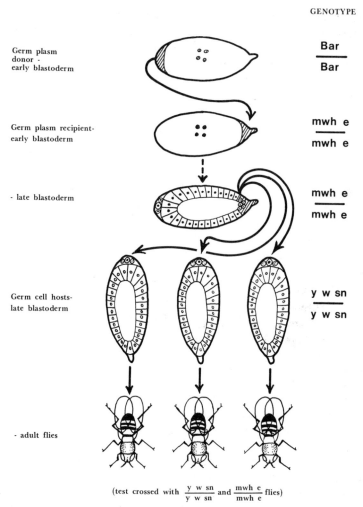

Germ plasm donor - early blastoderm — $\dfrac{\text{Bar}}{\text{Bar}}$

Germ plasm recipient- early blastoderm — $\dfrac{\text{mwh e}}{\text{mwh e}}$

- late blastoderm — $\dfrac{\text{mwh e}}{\text{mwh e}}$

Germ cell hosts- late blastoderm — $\dfrac{\text{y w sn}}{\text{y w sn}}$

- adult flies

(test crossed with $\dfrac{\text{y w sn}}{\text{y w sn}}$ and $\dfrac{\text{mwh e}}{\text{mwh e}}$ flies)

Figure 3.5 The germ plasm experiment in *Drosophila*. Some of the germ cell hosts are shown by test crosses to possess germ cells, induced by the germ plasm injection in the anterior pole of the recipients, of the genotype *mwh e/mwh e*. (Redrawn from Illmensee and Mahowald).

isolated from polar granules (Waring and colleagues, 1978). Following the suggestions of Mahowald (1968) this protein may represent the molecular site for the localization of a maternally-synthesized messenger RNA necessary for the production of a protein required for the formation of primordial germ cells. As with the amphibians, the germ plasm therefore acts as an autonomous, mosaic instruction.

In *Drosophila subobscura*, a maternal genetic effect similar to that of shell coiling in *Limnea* has been described. Female flies homozygous for the allele *grandchildless* (*gs/gs*) produce offspring which are sterile (Fielding, 1967). The effect is apparently autonomous to the ovaries, as reciprocal transplantation of larval gonads between *gs/gs* and normal flies does not affect the fertility of the gonads. In terms of the analyses of the properties of the germ plasm presented by Illmensee and Okada, it would then be expected that there is some deficiency in the action of the germ plasm formed in the oocytes of the *gs/gs* female. Gehring (1973) has reported that germ plasm is of normal histological appearance when examined with the electron microscope, but that the nuclei do not migrate into it in the normal way from the egg yolk. Instead, they move first to other regions of the egg cortex. As the effect is not dependent on the genotype of the nuclei, this indicates that their migration is directed by some feature of the regional organization of the egg determined by the maternal genotype. Interestingly, UV irradiation of the germ plasm of anuran amphibian eggs results in a rather similar delay in the migration of germ cells to the genital ridge (Züst and Dixon, 1977). This is thought to be due to irregular cleavage in the germ plasm region, and damage to a maternal component of the migratory mechanism (probably the mitochondria) in the germ cells.

Maternal genetic effects in *Drosophila melanogaster*

Several other loci are known to produce maternal effects in *Drosophila melanogaster*. Bull (1966) has shown that the *bicaudal* allele acts in this way to produce embryos and larvae with two abdomens (positioned anterior end to anterior end) and lacking heads. The reversed-position posterior ends appear to have normal spiracles, hindgut and Malpighian tubules. Similar larvae have been produced in the chironomid midge, *Smittia*, by UV irradiation of the anterior cytoplasm (Kalthoff and Sander, 1968). These defects could result from an abnormality in a reference point, or by a mosaic of head or abdomen determinants rearranging themselves symmetrically in the anterior and posterior halves of the egg. As it seems rather improbable that UV irradiation of a particular area could produce the latter effect, a more likely explanation of UV irradiation and the bicaudal locus is that they both affect a reference point. Kalthoff (1971) found that the effect of UV irradiation in *Smittia* could be reversed with visible light, and postulated that nucleic acids may be involved in the action of the postulated reference point. The more recent finding (Kandler-Singer and Kalthoff, 1976) that the application of ribonuclease to the anterior pole of *Smittia* eggs results in the formation of double abdomens seems to support this idea. In both cases the location and action of the germ plasm at the posterior pole is unaffected.

Several other maternal effect loci have been identified in *Drosophila melanogaster*. For example, Shannon (1972) found that homozygous *almondex* (*amx/amx*) females can produce heterozygous (*amx/+*) daughters which according to their own genotype should not be abnormal but which have a thorax with distorted body segments, bent wings and deformed or absent legs. A similar syndrome is produced by UV irradiation of the mid-lateral cytoplasm of the egg; in both cases the rest of the body appears normal. It is not clear whether a mosaic instruction has been disturbed or whether the action of a reference point has been disrupted.

Maternal effects in amphibians

The value of the Mexican axolotl, *Ambystoma mexicanum*, a neotonous salamander, as a genetic organism owes much to the work of R. R. Humphrey who, recognizing its potential for embryological studies, set out over 25 years ago to isolate mutant genes in this organism. Of the 20 or so mutant genes now available, a number act as maternal-effect lethals. For example, the f allele leads to the excessive accumulation of fluid in the blastocoel and the v allele to developmental arrest at the blastula stage (Malacinski and Brothers, 1974). The cl allele results in irregular cleavage, perhaps through disruption of the normal action of reference points for this process, whereas eggs derived from mothers homozygous for the nc mutant fail to cleave at all (Raff and Raff, 1978). Available evidence suggests that nc eggs lack organizer centres for the elaboration of microtubules.

The o allele has particularly interesting effects (Briggs and Cassens, 1966). Eggs from homozygous o/o females reach the blastula stage but fail to gastrulate (figure 3.6a,b). The increase in the rate of protein synthesis which is normally observed at the mid-blastula stage does not occur in these embryos. Injection of protein recovered from the germinal vesicle of immature eggs, or of cytoplasm of mature eggs (after the collapse of the germinal vesicle), carrying the normal allele (o^+/o^+ or o^+/o) will rescue the homozygous o/o embryos from these defects (figure 3.6c). It would appear therefore that the o^+ allele specifies a maternal protein which is necessary

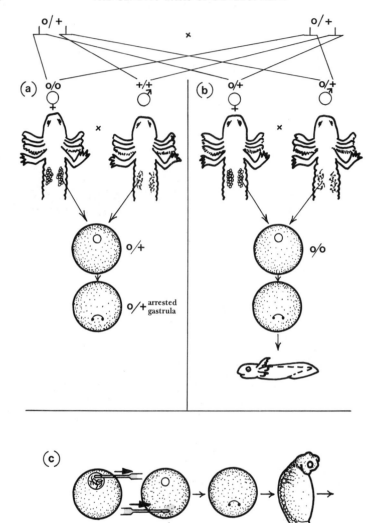

Figure 3.6 The maternal effect allele *o* (ova deficient) in the Mexican axolotl, *Ambystoma mexicanum*. (a) *o/+* progeny of an *o/o* female do not develop past the gastrula stage. (b) *o/o* progeny of an *o/+* female develop normally; comparison with (a) defines the effect of maternal genotype. (c) Injection of cytoplasm from the germinal vesicle of oocytes of *o/+* or *+/+* females into an egg from an *o/o* mother rescues the egg from the maternal effect. (d) Serial clonal nuclear transfers of activated nuclei into enucleate mutant eggs: a—nuclear

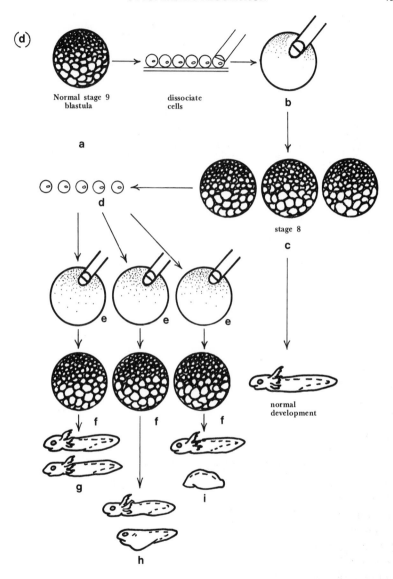

(d)

Normal stage 9
blastula

dissociate
cells

b

a

d

stage 8

c

e   e   e

normal
development

f   f   f

g

i

h

donor is a normal late blastula; b—original nuclear transfers; c—resulting stage 8 blastulae (some are allowed to develop and produce normal tadpoles); d—some are dissociated; e— these nuclei are used for the first serial transfer; f—resultant clone of blastulae; g, h, i—the type of development can be assessed and is often normal. (a)–(c) redrawn from Briggs and Justus; (d) redrawn from Brothers.

for the activation of nuclei and for normal development from the mid-blastula stage. Brothers (1976) demonstrated by nuclear transfer of genetically marked nuclei from normal embryos to enucleate eggs of o/o mothers that this activation is stable. Using this technique, she found that the normal nuclei became competent to gastrulate at the mid-blastula stage, and that the activation was stable when nuclei from such transplant embryos were serially transferred into additional enucleate eggs from o/o mothers (figure 3.6d). The mode of action of the o^+ protein remains to be elucidated, but it is clear that normal development depends on the interaction of this maternal protein with the nucleus of the developing egg.

The *deep orange* (*dor*) locus in *Drosophila melanogaster* provides an analogous situation. In addition to the defect in eye pigment synthesis, development is arrested during gastrulation in eggs of homozygous *dor/dor* mothers only. However, they may be rescued by the injection of cytoplasm from a $+/+$ egg (Garen and Gehring, 1972). Heterozygous $+/dor$ embryos are semi-viable; the paternal wild-type (*dor$^+$*) allele must be capable of providing the deficient cytoplasmic component before the critical stage. It can thus be seen that if the paternal genome is active sufficiently early, the effect of maternal-effect genes can be reversed.

Maternal effects in the mouse

A maternal-effect locus has been described in the mouse. The hairpintail allele (T^{hp}) is a deletion at the t locus of chromosome 17 (see chapter 8). Embryos carrying this mutation derived from the female parent show lethal abnormalities (including polydactyly and failure of the neural folds to close), but their littermates and all the progeny from the reciprocal cross (T^{hp} from the father) are viable (Johnson, 1974).

Transfer of eggs from normal females to $T^{hp}/+$ females, and vice versa, establishes that the effect is not due to uterine environment but to some effect of maternal genotype on egg cytoplasm. Winking (1981) recently investigated this very ingeniously by deriving a strain with a translocation of chromosome 17 so that oocytes diploid for chromosome 17 and heterozygous for $T^{hp}/+$ were produced. When fertilized by aneuploid sperm deficient for chromosome 17 from a male translocation carrier, diploid embryos resulted which were viable and normal (figure 3.7). The $+$ allele can therefore overcome the cytoplasmic effect of the T^{hp} allele if it is present during the haploid phase of the egg. A maternal effect mutant giving rise to tissue-specific defects might seem puzzling, since the mouse egg does not appear to contain instructions in either mosaic or reference

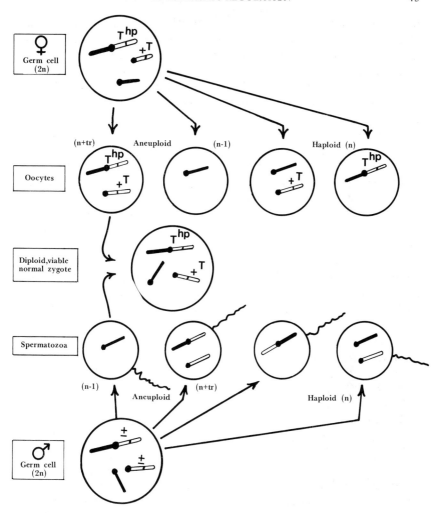

Figure 3.7 Chromosome segregation in carriers for a translocation of chromosome 17 (outlined) on to another autosome (solid). The translocated chromosome in the female parent carries the T^{hp} allele. (n = haploid complement, tr = translocated chromosome).

point form. However, the cytoplasmic defect could either block the normal development of such instructions or (as was suggested with *grandchildless* in *Drosophila*) it could prevent the expression of a reference point or mosaic instruction later in development. The crucial point is that in neither case would the cytoplasmic substance need to be localized in a particular region of the egg. A further possibility needs to be considered. There is now considerable evidence linking the t region of chromosome 17 with the expression of cell surface antigens restricted to the early stages of development (see chapter 8 for a more detailed discussion). The maternal effect of T^{hp} may arise from the expression of an altered cell surface protein.

A different type of maternal effect has been described by Wakasugi and Morita (1977). They found that female mice of the normally fertile DDK inbred strain became semi-sterile when mated to C57BL/6J males, as a result of a failure in trophoblast formation and embryonic death at the morula-blastocyst-egg cylinder stage of development. This abnormality was not seen in the reciprocal cross. This phenomenon could be due either to some incompatibility between the F1 genotype and the uterine environment of the DDK strain, or between the F1 nucleus and the DDK egg cytoplasm. DDK and C57BL ovaries were grafted into female mice of different genotype and then mated with C57BL and DDK males in all four combinations. A repeat of the original results was obtained; there is therefore no evidence for any effect of uterine environment. From this it was concluded that the DDK egg cytoplasm differs from that of C57BL, and that the C57BL component of the F1 nucleus cannot respond normally to the DDK cytoplasm. The genetic factors affecting the cytoplasm peculiar to the DDK strain, and the loci in the C57BL genome determining the inability to respond to it and form trophoblasts were not identified.

Colaianne and Bell (1970) have found a rather similar interaction in *Drosophila* between the product of the sonless (*snl*) locus and the Y-chromosome, resulting in the absence of male offspring in crosses between homozygous *snl/snl* females and normal (snl^+/Y) males.

Cortical inheritance
Several observations suggest that the cell membrane and the stiff cortical layer immediately beneath it are very important as sites of cytoplasmic determinants. For example, in ascidians such as the marine tunicate, *Ciona*, normal development (including the appearance of tissue-specific enzymes) will still occur if the cytoplasm is disorganized by centrifugation or even if up to half of the cytoplasm is surgically removed (Conklin, 1931). Cortical localization of morphogenetic determinants has also been demonstrated in the squid egg by Arnold (1969). UV irradiation of small areas of the thin cortical region of the egg before cellularization does not affect cleavage, and the site of irradiation becomes included in the developing

blastoderm. However, the embryo lacks specific organs and tissues normally developing from the irradiated site. UV irradiation (Nöthiger and Strub, 1972) and more recently laser microbeam destruction of blastomeres (Lohr-Schardin and colleagues, 1979) has been similarly used to produce detailed fate maps of the surface blastoderm of the developing insect egg.

Curtis (1960, 1962) carried out some elegant experiments involving cortical grafting in *Xenopus laevis*, and found that the cortex of the grey crescent region is responsible for the subsequent formation of the blastopore organizer (which in a sense programmes the rest of development through the mechanism of primary induction) (figure 3.3). Recently, however, Scharf and Gerhart (1980) have shown that the organizer activity of the grey crescent region may be transferred to other regions of the egg by so-called oblique orientation of the egg (holding the egg with the animal-vegetal pole axis at 90° to the vertical). This treatment also rescues eggs from the detrimental effect of UV irradiation of vegetal hemisphere on organizer activity. Although the effect of UV light is unknown, it is perhaps significant that its effects can be mimicked in squid eggs by the topical application of cytochalasin B. This drug has been implicated as an inhibitor of contractile microfilaments. The effects of both treatments may therefore be to alter the motile machinery of the egg cortex. The effect of oblique orientation of UV irradiated eggs may be to promote the internal rearrangements which would normally occur before the first cleavage division and lead to the establishment of organizer activity in the region of the grey crescent.

Little is known about the establishment of cortical determinants in multicellular organisms. Although the nucleus of the egg or the attendant nurse cells in the ovary could be involved, the observations of Sonneborn (1970) on the inheritance of abnormal forms of *Paramecium* suggest that cortical inheritance could be dependent on the physical continuity of the cortex through the germ cells to the egg. Sonneborn found that some clones of *Paramecium* developed abnormal patterns of cilia on their surface. The rows of cilia in a particular region might be orientated in an unusual way; occasionally a double ("Siamese twin") form would be found. When such *Paramecia* were conjugated with normal forms so that nuclear and cytoplasmic exchange took place, these exchanges had no influence on the form of descendants. The clones continued to show stable and uniform inheritance of their unusual morphology (figure 3.8). The basis for this unusual mode of inheritance is not understood, but these observations could be relevant to the determination of egg structure.

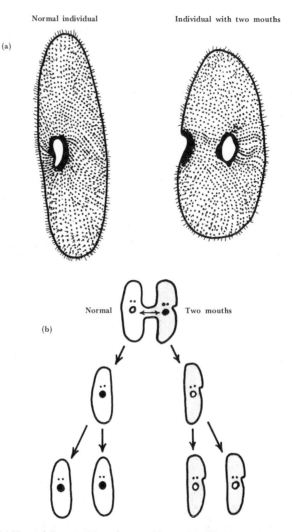

Figure 3.8   (a) Normal *Paramecium* and unusual form. (Modified from Sonneborn).
  (b) The consequence of conjugation (with nuclear and cytoplasmic exchange) between the two forms.

Nuclear-cytoplasmic interactions and cell fusion experiments

The earlier sections of this chapter describe convincing evidence for the establishment of egg structure and cytoplasmic determinants as a result of

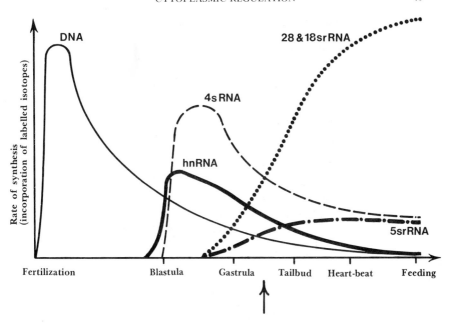

Figure 3.9 Cytoplasmic control of nucleic acid synthesis in *Xenopus*. The figure shows the sequence of nucleic acid synthesis in normal development. When a neurula nucleus engaged in the synthesis of all classes of RNA (arrowed) is transplanted to a fertilized egg, RNA synthesis virtually ceases within 20 minutes. As the recipient egg develops, the normal pattern of synthesis is seen. (Experiments of Gurdon and Woodland).

the activity of the maternal genotype. In this section, some experiments which demonstrate directly that the cytoplasm has metabolic effects on the nucleus will be described.

It is implicit in the interpretation of the nuclear transfer experiments in amphibians in the last chapter that the nuclei of differentiated cells respond to the cytoplasm of the egg in which they find themselves. As the different stage nuclei synthesize the different classes of nucleic acid in characteristic proportions (figure 3.9) this supposition can be tested experimentally. In fact, it is found that shortly after transfer, the nucleus changes its pattern of synthesis to that of the host cytoplasm (including the increased synthesis of some types of nucleic acid, and reduced synthesis of others). The pattern of synthesis follows the normal course of development thereafter.

Cell fusion is a technique which enables the study of the effects on gene activity of the cytoplasm of differentiated cells. Viral infection often results in the fusion of neighbouring cells into a large multinucleate cell. Harris and Watkins (1965) showed that UV-inactivated Sendai virus particles could cause the fusion of cells of different types and different species in culture. The fused cells are viable and can be examined cytologically. The synthesis of DNA and RNA can also be determined by employing the technique of autoradiography to follow the incorporation of tritiated

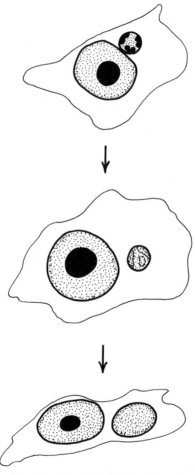

Figure 3.10   Sequence of events after the fusion of a HeLa cell and a hen erythrocyte. The erythrocyte nucleus (right) swells considerably and loses the deeply staining nuclear bodies (heterochromatin). (Redrawn from Harris).

thymidine or uridine in the nucleus. The interaction of nucleus and cytoplasm can be most readily examined in fusions between cells containing relatively inactive and active nuclei. For example, fusion of HeLa cells (a human tumour line with an active nucleus) with hen erythrocytes (with an inactive nucleus) results in a hybrid cell. Since Sendai virus lyses the erythrocyte during fusion, the cytoplasm of the hybrid is largely of HeLa origin. Following fusion, the erythrocyte nucleus enlarges considerably and deeply staining nuclear bodies (representing synthetically inactive heterochromatin) gradually disappear (figure 3.10). These changes are accompanied by an increase in biosynthetic activity in the nucleus. As similar changes are not seen when two different inactive cell types are fused, they cannot be directly due to the action of the virus. The HeLa cytoplasm must contain substance(s) which stimulate RNA synthesis in the nucleus.

These two systems have been widely used in studies of the regulation of gene expression. Experiments with labelled proteins show that they move between the cytoplasm and nucleus of the amphibian oocytes (Ecker and Smith, 1971), but the nature and mode of action of cytoplasmic signals remain to be elucidated. Specific inducing agents including hormones, are discussed in more detail in chapters 4 and 5, and the use of cell hybrids (involving nuclear as well as cellular fusion) is referred to again in chapter 6.

Conclusions

Wilson, in 1925, wrote that "the cytoplasmic organization of the egg is itself the product of an antecedent process of epigenetic development in the course of which, as we have every reason to believe, the chromosomes have played their part...the chromosomes are as much concerned in the determination of the so-called 'preformed' or cytoplasmic characters as in any other". At the same time "heredity is effected by the transmission of a nuclear preformation which in the course of development finds expression in a process of cytoplasmic epigenesis". Separate and non-interacting roles for the nucleus and cytoplasm were rejected. The evidence presented in this chapter provides strong confirmation of this viewpoint, and an initial model of the developmental process can be described.

In chapter 2 it was concluded that the nuclei of differentiated cells are totipotent in most cases, and differences between their genetic complements cannot therefore be implicated as the origin of differences between cells. To the extent that chromosomal differentiation does occur it is characteristic of particular specialized cell types. Cytoplasmic determinants in different regions of the egg must therefore be postulated and the experiments described in this chapter provide compelling evidence for their existence. Furthermore, normal development requires that the nucleus is genetically competent and responds to the cytoplasm, as described in the cases of sterility in the mouse and the *o* locus in the axolotl. Finally, evidence is presented of changes in nuclear activity taking place under the influence of the cytoplasm during the development of the amphibian egg and in differentiated cells in culture. This model of development therefore predicts that nuclei in different regions of the egg and different tissues should possess different patterns of nuclear activity in response to the cytoplasmic determinants, such that the initial cytoplasmic difference would be amplified by the differential appearance of gene products. These gene products could then affect the nuclear activity of the cell or of other cells to which it may have access, and in animals they could control the movement of the cell by affecting its motility and the composition of the cell surface. Each cell type would undergo a characteristic pattern of differential gene activity with time. In the next three chapters, evidence concerning differential gene function and the mechanisms involved will be discussed. Other topics are taken up in succeeding chapters.

The nature of the cytoplasmic determinants has been clarified by the studies described, and it seems possible that at last some of them will soon be identified. The germ plasm of insects and anurans seems to be of the mosaic type, but there is probably no germ plasm in this sense in reptiles, birds and mammals. The bicaudal mutation and the UV- and ribonuclease-induced phenocopies of it seem to establish a reference point (which could act by means of a chemical gradient, or some other signal), and the developmental information localized in the cortical region of eggs of *Ciona*, squid and frog may likewise depend on the arrangement of reference points. Many of the maternal-effect mutants described in this chapter have general effects on the egg and probably depend on the absence of some cytoplasmic factor required for normal cellular function and behaviour. The developmental arrest seen in *dor* in *Drosophila melanogaster* and in *o* and *nc* in the axolotl would appear to be of this type. Apart, therefore, from *gs* in *Drosophila subobscura* and its relation to the mosaic germ plasm, there is no case where a mosaic instruction can be unequivocally shown to be involved. Finally, in the early mouse embryo, no evidence either for a mosaic or for fixed reference points can be found; the blastomeres are apparently totipotent and the developmental fate of cells depends on their final position in the embryo. For a maternal-effect mutant to be expressed, a maternally-determined component of the egg must have developmental significance. T^{hp} in the mouse may depend therefore on a maternally-determined developmental instruction that is essentially labile during early development but which eventually becomes fixed in a particular region of the embryo.

CHAPTER FOUR

THE MOLECULAR BIOLOGY
OF CHROMATIN AND NUCLEIC ACIDS

Development proceeds by means of interactions between the nucleus and the cytoplasm, leading to a specific pattern of gene expression in each cell type during its cytodifferentiation, as discussed in the previous chapters. Thus the composition of the cell is progressively changed. Some evidence that cytoplasmic factors can influence gene expression by effects on chromosome condensation and transcription was described. The purpose of the succeeding chapters is to investigate the mechanisms involved in the regulation of gene expression. As background information to the genetic studies of regulation described later, this chapter provides a brief intro-duction to biochemical approaches to the eukaryotic cell, including the recently developed techniques of restriction mapping and genetic engineer-ing. A much fuller account of the molecular biology of eukaryotic cells can be found in Hood, Wilson and Wood (1974) and Lewin (1980). For a discussion of the biochemical approach to embryology, see Davidson (1976).

The eukaryotic gene-action system
The eukaryotic gene-action system for the expression of coding sequences (as presently understood) is summarized in figure 4.1. DNA in the chromosome is transcribed to complementary, primary transcript RNA, also called heterogeneous nuclear (hn) RNA. This hnRNA is modified within the nucleus by cutting and subsequent splicing together of RNA fragments. Many of the RNA fragments remain in the nucleus and are eventually degraded there. The RNA which is to be transported to the cytoplasm is polyadenylated. The polyadenylated, cut and spliced RNA can then serve as cytoplasmic messenger (mRNA). The mRNA is trans-

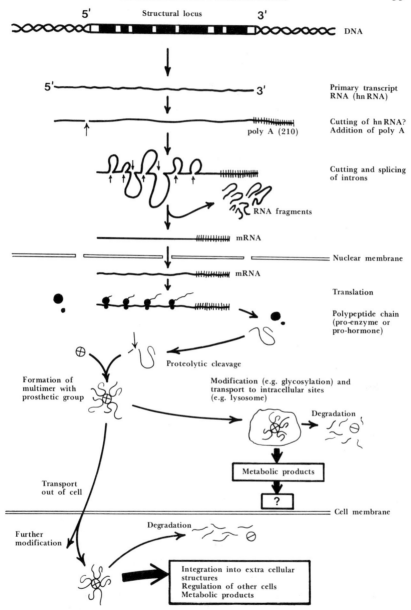

Figure 4.1 Some features of eukaryotic gene action systems.

lated into the corresponding amino acid sequence on the ribosomes. Depending on the particular polypeptide thus produced, its fate is then quite variable. It may immediately function as an enzyme structural protein, but commonly it will be broken down by specific proteolysis, so that a pro-hormone or pro-enzyme gives rise to an active molecule. It may be glycosylated or otherwise changed. Often it will aggregate with similar or different types of polypeptide chain to form an oligomer, with or without a prosthetic group. It may be transported to specific sites within the cell, or outside the cell. It can then exert metabolic effects in conjunction with the other constituents of the cell or in target cells. Ultimately it will be degraded, whether inside or outside the cell. Other categories of genes are expressed in a different manner. rRNA and tRNA sequences are of course specified directly by the DNA sequences. Regulatory sequences and controlling elements (chapter 6) may act by affecting the process of transcription.

It is a frequently used convention to regard the gene action system as ending with the production of functional protein (or enzyme). Whilst this is convenient, it is also arbitrary, since in eukaryotes other gene-action systems can affect the process of expression before the functional protein has been produced, as well as afterwards (e.g. in a metabolic pathway). In this and the next chapter, regulation of metabolism within the cell is considered. One gene-action system (in this sense) can be regulated by effects of other gene-action systems at any level from chromosome organization, through transcription, nuclear RNA processing and translation to the degradation of functional protein. The consideration of gene interactions and the co-ordination of gene expression within the cell (chapter 6) is a natural extension of this approach. Later chapters consider gene-action systems where the phenotype depends on interactions between cells.

This chapter deals initially with the structure of chromatin and functional changes at the chromosome level, proceeds to a brief description of the kind of fine-structure mapping of chromatin which is permitted by the technology of genetic engineering and concludes with a description of some systems where the biochemical mechanisms for regulating transcription are at least partly understood.

Chromatin fibres and nucleosomes
The material of the chromosomes (known as chromatin) contains DNA, the basic histone proteins rich in arginine and lysine, non-histone (non-

basic) proteins and some RNA (Evans, 1977). The structural continuity of the chromosome is provided by the DNA molecule, since DNAase breaks the chromosome down, but RNAase or proteases do not. In confirmation of this, four molecules of DNA of differing molecular weights corresponding to the lengths of each of the four haploid chromosomes can be isolated in *Drosophila* (Kavenoff and Zimm, 1973). A naked DNA double helix has a diameter of about 2.2 nm; fibrils about 3.0 nm diameter have often been reported by electron microscopists (e.g. Solari, 1971) so this probably represents the basic DNA duplex and associated material. Larger fibrils with diameters of 10 nm can also be seen, and X-ray diffraction studies indicate that these could consist of coiled fibrils of the first type with a pitch of about 11 nm (Pardon and colleagues, 1973). Enzymatic digestion with various nucleases tends to produce fragments of DNA about 200 base pairs long, suggesting that much but not all of the DNA in chromatin is present in repeated structural units of some kind (Hewish and Burgoyne, 1973).

The histone proteins can be divided into five groups on the basis of their arginine or lysine content and chromatographic behaviour on gel filtrations; these are known as H1 (lysine-rich), H2A and H2B (slightly lysine-rich) and H3 and H4 (arginine-rich) (Bradbury, 1975). These proteins show little variation between tissues and their sequences (apart from H1) are strictly conserved over a variety of species, whereas the non-histone proteins are much more variable in both respects. The histones H2A, H2B, H3 and H4 are also found in roughly equal molecular proportions and in constant stoichiometric proportion to DNA. This suggests that they are important in the structure of chromatin. Kornberg and Thomas (1974) found that mild fractionation procedures yielded tetramers with two molecules each of H3 and H4 and H2A-H2B oligomers. These would not associate spontaneously, but in the presence of DNA duplex strands they reacted to give a material with the same X-ray diffraction properties as chromatin. Histone aggregations containing eight molecules can be isolated from chromatin, so Kornberg (1974) proposed that the structure of chromatin is based on the continuous DNA duplex with repeating units of the histone octamers. Since eight molecules of histone are found for each 200 or so base pairs of DNA, the digestion of DNA by nucleases into 200 base-pair fragments could be explained if the DNA is associated with the repeated histone unit in such a way that it is protected from digestion except for the region between units.

Various electron microscopists, following Olins and Olins (1974) have found that chromatin preparations give rise to pictures howing spherical bodies about 6–10 nm in diameter connected by a thin fibril corresponding to the DNA duplex. These bodies, now known as *nucleosomes*, are produced in the reaggregation experiments with histone octamers and DNA (Oudet and colleagues, 1975). Analysis of isolated nucleosomes shows the presence of the four histones, so that they are generally accepted as a basic feature of chromatin organization.

How is the DNA duplex associated with the nucleosome particles? Neutron diffraction allows discrimination between DNA and protein, and

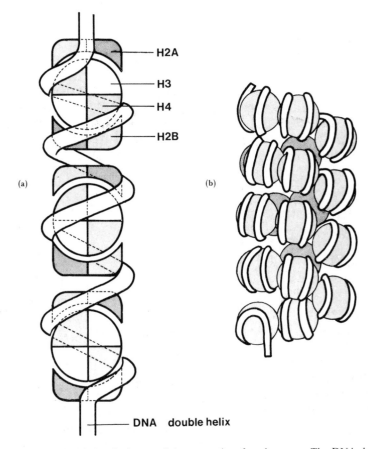

(a)

(b)

H2A

H3

H4

H2B

DNA double helix

Figure 4.2 (a) Model showing some of the properties of nucleosomes. The DNA double helix is wound around the outside of a protein core made up of two molecules of each of four species of histone. (b) Diagram of the proposed arrangement of nucleosomes in a solenoid. (Modified from Evans).

the diffraction pattern for DNA gives a wider radius than that for protein (Pardon and colleagues, 1975). One model of the structure of the nucleosomes is summarized in figure 4.2a. 200 base pairs would code for $\frac{200}{3} = 67$ amino acids. During transcription of a structural gene, the DNA from several nucleosomes would therefore need to be traversed by the RNA polymerase enzyme. Recent evidence suggests that the DNA-nucleosome complex is broken down during transcription (Weintraub and Groudine, 1976; Wu and colleagues, 1979).

Chromosome coiling and supercoiling

Histone H1 is found in lesser amounts than the other histones. It is present in greater amount in condensed (coiled) chromatin, which stains readily and is known as *heterochromatin*, than in soluble chromatin or *euchromatin*. In the absence of H1, 146 base pairs of DNA are coiled around the central histone octamer. H1 binds at the site of DNA entry and exit to the octamer in association with a further 20 base pairs of DNA to give two complete super-helical turns around the nucleosome core (Laskey and Earnshaw, 1980). In heterochromatin, the nucleosomes are stacked in the form of a solenoid (figure 4.2b). So far at least three degrees of coiling have been described: the basic DNA double helix, the coiling of this round the nucleosomes, and the association of the nucleosomes into a solenoid. From the length of human chromosomes during metaphase of cell division where DNA is contracted to a fraction of 5×10^{-3} of its uncoiled lengths, there must also be at least a fourth degree of coiling of the chromatin fibre (Bahr, 1970). Models for the third and fourth and higher degrees of coiling are being elaborated, but are still speculative (e.g. Worcel and Benyajati, 1977). It seems likely that the non-histone proteins are also involved in some of the degrees of supercoiling.

The functional significance of heterochromatin is discussed later.

Chromosomal organization

Several techniques have recently been developed which stain the chromosome in a number of transverse bands. Pardue and Gall (1970) showed that staining with Giemsa following alkali denaturation of DNA accentuated the heterochromatic centromeric regions of the chromosomes (figure 4.3), and that these so-called "C-bands" contained the satellite (highly-repetitive) DNA, as detected by *in situ* hybridization with labelled RNA complementary to satellite DNA. The other regions of the chromo-

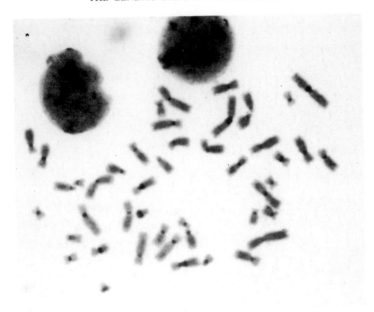

Figure 4.3   Banding patterns of human metaphase chromosomes—C-banding. (Photograph kindly provided by Dr. A. C. Chandley).

some are barely stained following denaturation. (Bands can also be found in some non-centromeric regions, for example, in the human Y-chromosome). Less severe treatment (e.g. with trypsin) or use of other stains (such as quinacrines which are UV fluorescent) gives rise to different sets of bands, named G- and Q-bands respectively—figure 4.4. Many variations of these techniques have been developed (Evans, 1977). In man, about 300 bands can be seen at metaphase, but these are formed from a larger number (perhaps 3000) which can be seen in the less-contracted prophase chromosomes.

The mechanism behind these banding phenomena is controversial (see Evans, 1977). It seems likely that C-banding is due to the tight coiling of the relevant regions and consequent resistance to denaturation. For present purposes, it is important only to note (a) that these staining methods reveal some type of structural organization in the chromosome, and (b) banding techniques have enabled separate identification of chromosomes, or fragments of chromosomes, which would otherwise appear identical, and they have therefore been of the utmost utility in cytogenetic experimentation (Miller and colleagues, 1973).

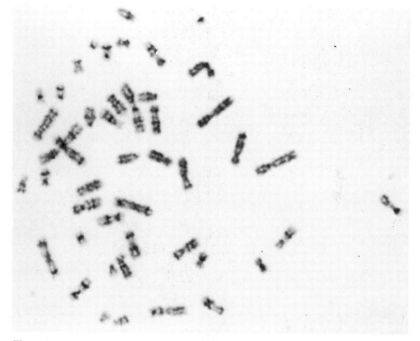

Figure 4.4 Banding patterns of human metaphase chromosomes—G-banding. (Photograph kindly provided by Dr. A. C. Chandley).

Chromosome coiling does not remain constant during the cell-cycle. The compact coils formed during cell division have the physical function of allowing the centromeres to distribute the chromosomes appropriately to the daughter cells. Apart from this, the coiling of chromosomes could have functional consequences. There are several experimental systems where the uncoiling of chromosomes is associated with transcription. The puffing of polytene chromosomes and lampbrush chromosomes have each been studied by a combination of cytological and biochemical techniques, summarized briefly below.

Puffing of polytene chromosomes

During development, the bands of polytene chromosomes in *Drosophila* undergo reversible morphological changes, notably, swelling and apparent unfolding of the chromosomal strands—figure 4.5. The swollen bands are known as puffs or Balbiani rings (Beerman, 1963, 1964). Judging by the

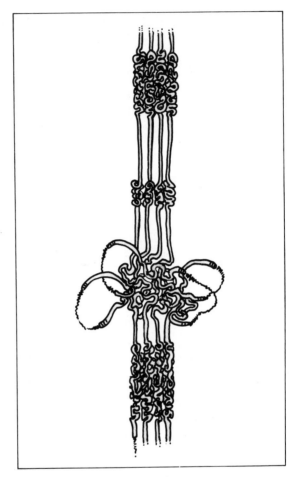

Figure 4.5 Diagram of the structure of a puff in a polytene chromosome, showing the unravelling and uncoiling of the chromatids. (Redrawn from DuPraw).

effects of antibiotics which inhibit transcription, the base composition of RNA isolated from puffs, and the incorporation of tritiated uridine as determined by autoradiography, puffs are visible signs of increased transcription of the relevant loci. Relatively few of the bands puff, and these show patterns of activity specific both for developmental stage and tissue (figure 4.6). This system therefore provides good evidence for the postulate of transcriptional regulation of gene expression during development.

In a particularly elegant series of experiments, Doane (1969, 1971, 1975)

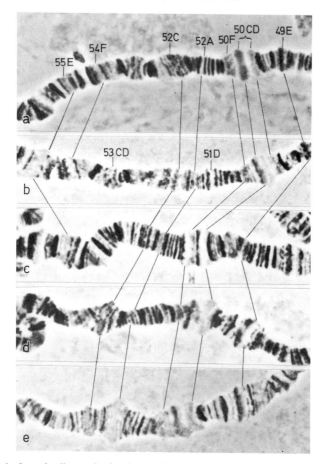

Figure 4.6 Larval salivary gland polytene chromosomes from *Drosophila melanogaster*. A series of photographs of a section of chromosome arm 2R showing changes in puffing activity during development. (a), 110-h old larva, (b) 115 h larva, (c) 0 h prepupa, (d) 6 h prepupa and (e) 8 h prepupa. (From M. Ashburner (1972) in *Results and Problems in Cell Differentiation*, Vol. 4, (W. Beermann, ed.), Springer-Verlag, Berlin.

was able to show that puffing of the cytogenetic locus coding for amylase in *Drosophila hydei* correlated with the observed enzyme activity in different tissues at various developmental stages. The functional significance of puffing is shown by the fact that the addition of starch to the diet increased puffing and also amylase levels in the midgut.

The developmental significance of puffing can be illustrated by looking

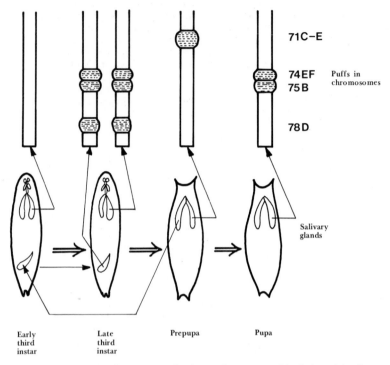

Figure 4.7   The normal puffing pattern of polytene chromosome 3 in the larval development of *Drosophila*, and its recapitulation in glands transplanted from the prepupa to the early third instar larva. (Redrawn from Suzuki).

at the factors which cause puffing, and by examining the biochemical consequences of puffs. Salivary gland chromosomes transplanted from a prepupa to the early third instar quickly lose their characteristic puffs (band 71 C–E in figure 4.7), and acquire the puffs appropriate to their host's developmental stage. Kroeger has carried out a similar experiment where he transplanted nuclei from larval salivary glands into pre-blastoderm and blastoderm stage embryos, and found that the host cytoplasm altered the puffing pattern (see Fischberg and Blackler, 1961).

A clue to the factor which induces the puffing pattern during pupation is provided by experiment where larvae are ligatured at the third instar (figure 4.8). The posterior half of the larva is not exposed to ecdysone (moulting hormone) from the ring gland and the puffing remains larval in type, whilst the anterior part of the larva pupates and the chromo-somes display the normal puffing pattern. Clever and Karlson (1960)

Table 4.1 Target sequences of some useful restriction endonucleases. Note that in all these cases, the complementary sequences (read $5'-3'$) are identical and will therefore be cut in the same place (indicated by the arrow). If the arrow is not in the centre of the sequence the staggered cut will produce a characteristic single-stranded sequence protruding from the helix after the cut is made. This will hybridize with any other piece of DNA cleaved by the same enzyme, joining the two together, end to end, and is therefore known as a sticky end.

Source of enzyme	Abbreviation	Target sequence
		$(5'-3',$ $3'-5')$
		$\downarrow$
Anabaena subcylindrica	AsuI	GGNCC
		CCNGG
		$\uparrow$
		$\downarrow$
Bacillus amyloliquefaciens	BamHI	GGATCC
		$\uparrow$
		CCTAGG
		$\downarrow$
Bacillus globigii	BglII	AGATCT
		TCTAGA
		$\uparrow$
		$\downarrow$
Escherichia coli/RI	EcoRI	GAATTC
		$\uparrow$
		CTTAAG
		$\downarrow$
Haemophilas influentae(d)	Hind III	AAGCTT
		TTCGAA
		$\uparrow$
		$\downarrow$
Providencia stuarti 164	PstI	CTGCAG
		GACGTC
		$\uparrow$
		$\downarrow$
Xanthomonas badrii	XbaI	TCTAGA
		AGATCT
		$\uparrow$

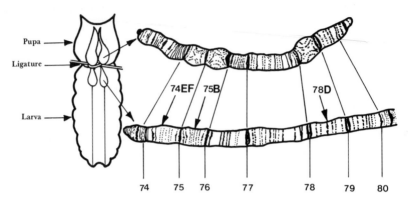

Figure 4.8 A larva is ligated so that anterior pupates while the posterior remains larval. The puffs in polytene chromosome 3 of the salivary glands are shown. (Redrawn from Suzuki).

showed, by injecting ecdysone into young fourth instar larvae of *Chironomus tentans*, that the changes in puffing pattern (like moulting and metamorphosis) are controlled by this hormone. Karlson (1976) also showed, by pulse labelling of RNA, that the synthesis of the enzyme DOPA decarboxylase (required for the synthesis of the *N*-acetyldopamine which is responsible for the tanning of cuticle) follows from the transcription of newly synthesized mRNA induced by ecdysone in *Calliphora* larvae. However, the function of the majority of the induced loci is unknown.

The complex mechanisms which co-ordinate the puffing pattern are discussed later in this chapter. The events leading up to the interaction of a steroid hormone (such as ecdysone) with chromatin have been especially well studied in vertebrates, and are also considered later in the chapter. Recently, cytoplasmic receptors, similar to those found in vertebrate tissues, which specifically bind ecdysone with high affinity and transport it to the nucleus, have been found using a highly radioactive ecdysone analogue, ponasterone A, prepared from a plant ecdysteroid, stachysterone C (Ashburner, 1980).

Lampbrush chromosomes
Chromosomes with a characteristic morphology are found in the oocytes of many vertebrates during the diplotene stages of meiosis; similar structures, supposedly resembling the brushes used to clean oil lamps in the last century are also found in the Y-chromosomes of *Drosophila* spermatocytes. Their relatively large size in amphibians facilitates dis-

section out of the ·nucleus and subsequent analysis by nucleic acid hybridization, or incorporation of labelled precursor followed by auto-radiography. A fuller discussion is to be found in chapter 8 of Davidson (1976).

The structure of a lampbrush chromosome is shown diagrammatically in figure 4.9; it consists of the two paired homologous chromosomes, each composed of two chromatids. The lateral loops consist of very thin filaments about 3–5 nm wide with some parts up to 10 nm wide; the material between the chromomeres consists of two fibres each about 10 nm wide. The loops probably represent uncoiled chromosomal material (5 % of the total), whilst the axis is coiled (95 %). Autoradiography shows that ^{3}H-uridine is incorporated into RNA along the loops, but not along the axes. Sequential incorporation of ^{3}H-uridine suggests that the loop may be in movement, being spun out from one end and wound up at the other so that longer regions of the axis are eventually transcribed. Transcription of the loops has been visualized by Miller and colleagues (1970) in the electron microscope. The very strong synthetic activity of the uncoiled loops is probably responsible for the formation of maternal cytoplasmic mRNA in the oocyte (about 1.5 % of the unique sequences).

In *Drosophila hydei* 5 loops can be seen on the Y-chromosome of meiotic prophase preparations from diploid spermatocytes. Analysis of flies with deletions for varying portions of the Y-chromosome reveals that all five loops are necessary for the differentiation of functional spermatozoa; deficiency results in shortened tails, abnormal mitochondria, or malforma-tion of the tail fibres (Hess, 1970). In all somatic tissue, the Y-chromosome is apparently inactive (XO flies are morphologically normal males), and is heterochromatic (deeply staining and tightly-coiled) so that the cor-relation between uncoiling of the chromosome and genetic activity is very clear in this case. Genetic studies of heterochromatin are described in the next chapter.

The technology of genetic engineering

The analyses of chromosome structure described above are leading to an understanding of the biochemical factors involved in chromosome coiling which may be important in the regulation of transcription of genetic loci.

With the application of several new techniques, the knowledge of nucleotide sequences of eukaryotic genes is accumulating at a very rapid rate. In addition to nucleic acid hybridization, the important developments are restriction mapping, gene cloning (genetic engineering) and rapid base sequencing methodology. Together, they permit fine-structure analysis of eukaryotic genes which was not possible five years ago. Here, we indicate the nature of the technology involved and some of the ways in which it is

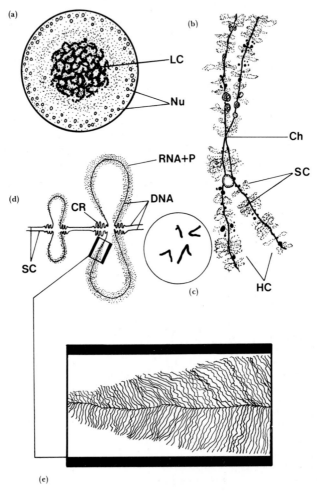

Figure 4.9 Lampbrush chromosomes of *Xenopus:*
  (a) Nucleus of maturing oocyte. LC—lampbrush chromosomes; Nu—nucleoli.
  (b) Two paired homologous chromosomes (HC). Ch—chiasma; SC—sister chromatids.
  (c) Normal mitotic chromosomes for comparison.
  (d) Diagram of structure. CR—chromomere; RNA + P—ribonucleoprotein.
  (e) Diagram of electron micrograph of loop region, showing transcribed ribonucleoprotein fibrids of increasing length. (a)–(c) Redrawn from Callan; (d) redrawn from Hadorn; (e) based on Miller.

being applied to the analysis of the transcriptional processes, and their regulation.

The principle of restriction mapping

Restriction mapping requires the use of bacterial endonucleases involved in the phenomenon of host-controlled restriction (modification) of bacteriophage infection. Briefly, phage grown on one strain will generally not form plaques on bacteria of a different strain. The bacteria possess endonucleases (known as restriction enzymes) which recognize and cleave characteristic and strain-specific short DNA base sequences of 4–7 base pairs. The bacterial DNA is immune to this action because of an enzymatic modification which protects it from restriction (Arber, 1974). The modification enzyme may also occasionally protect phage DNA before it is degraded, in which case the phage succeeds in growing in its new host. These restriction enzymes can be isolated from different species and strains

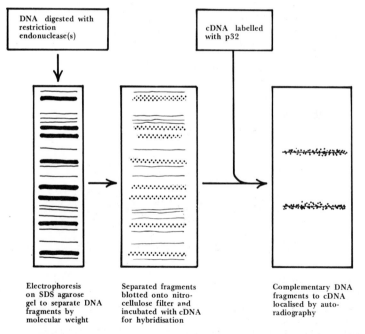

| DNA digested with restriction endonuclease(s) | | cDNA labelled with p32 |

| Electrophoresis on SDS agarose gel to separate DNA fragments by molecular weight | Separated fragments blotted onto nitro-cellulose filter and incubated with cDNA for hybridisation | Complementary DNA fragments to cDNA localised by auto-radiography |

Figure 4.10 The principle of the Southern blotting technique to identify which fragments of DNA produced by restriction enzymes are complementary to a particular cDNA preparation.

of bacteria, and used to cleave eukaryotic DNA preparations reproducibly at their specific sites (table 4.1, see also Murray, 1978).

In eukaryotic restriction mapping, such DNA fragments, representing perhaps 10^6 different DNA sequences from the source material, can then be separated by electrophoresis in a gel matrix (e.g. agarose) according to molecular weight. Simultaneously, a specific probe for the desired structural gene is prepared. For example, mRNA for a β-globin chain can be isolated at high purity, copied into cDNA by reverse transcriptase, and labelled with ^{32}P to high specific activity. This cDNA could also be cloned if large quantities were required—see below. This labelling of the DNA is achieved by a process called *nick translation*. Nicks are first introduced into the DNA by very limited exposure to DNAase I. At those nicks which expose a free 3'-OH group, DNA polymerase I of *E. coli* will incorporate nucleotides successively. The exonucleolytic activity of this enzyme will also mean that the nick will migrate ("translate") in a 5' to 3' direction. With the provision of the four ^{32}P-labelled deoxynucleotide triphosphates, the excision and addition of nucleotides will proceed along one strand of the DNA duplex to produce a highly radioactive but otherwise unchanged DNA molecule. The position of the DNA fragment(s) containing the coding sequence for β-globin in the agarose gel can now be determined by blotting the separated fragments onto a nitrocellulose filter, hybridizing with the probe, and locating the site of hybridization by autoradiography. This technique is known as Southern blotting (figure 4.10).

The restriction map is now constructed by comparing the results obtained with a variety of restriction enzymes (figure 4.11). For example, the enzyme EcoR I (named from the producing bacteria, *E. coli*) produces two β-globin hybridization bands on the gel. It must therefore cut the structural locus into two. The enzyme Pst I produces only one band, and so does not cut the locus (it does however, separate the β and δ chain loci). Used in combination (double digest), Pst I cuts one of the EcoR I fragments outside the coding sequence (seen by a reduction of its molecular weight on the gel) but not the other. In this way, a map of the sequences recognized by the various restriction enzymes can be constructed, and the distance apart of the cutting sites can be deduced from the size of the fragments produced with a variety of restriction enzymes. It also becomes evident that there are spacer sequences between the globin β and δ loci, as there are several different cutting sites available which do not disrupt the loci themselves.

Restriction mapping is a technique which can be applied to any locus for which a labelled probe sequence can be prepared. The probe does not have

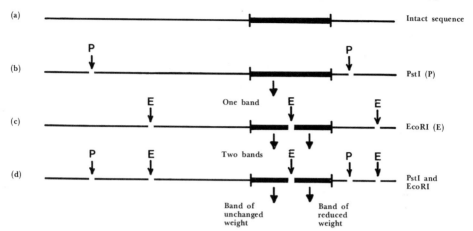

Figure 4.11 The principle of restriction mapping is shown in this simple example. The intact sequence (a) is cut outside the coding sequence (dark) by Pst I to yield a single hybridization band with Southern blotting (b). In contrast, Eco RI (c) yields two bands and therefore cuts within the coding sequence. When both enzymes are used together (d), one of these bands is unchanged in molecular weight so Pst I is cutting this molecule outside Eco RI. The other band is of reduced molecular weight, so Pst I is cutting it inside EcoRI. Relative distances are estimated by comparing the molecular weights of the fragments, and the map can be elaborated by the use of further restriction endonucleases.

to be pure so long as one species is present at several times the frequency of the next most common species. Relatively crude mRNA preparations from differentiated cells may therefore be used to prepare probes. Methods for obtaining a mRNA species include physical separation of mRNA's, especially from specialized cells elaborating large quantities of particular proteins, and selection of ribosomes producing the desired polypeptide by complexing with a specific antibody which will have the appropriate messengers attached. Reverse transcriptase produces single-stranded copy DNA (cDNA) which can be converted to double-stranded cDNA by the use of polymerase I. The cDNA can be used directly as a probe, or cloned to produce large amounts of cDNA with high specific activity.

Restriction mapping will also allow population studies at the DNA level, because polymorphisms may often be found within the population whereby a restriction site is present in some chromosomes and not within others as a result of mutation during evolution (e.g. rDNA in *Drosophila*, Wellauer and David, 1977; globins in man, Kan and colleagues, 1980).

Gene cloning

The technique of gene cloning consists of bringing a specific piece of DNA

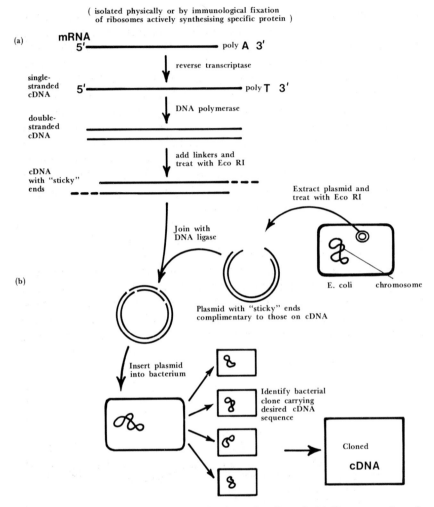

Figure 4.12 The principles of DNA cloning in *Escherichia coli*. (a) The preparation of double-stranded cDNA for insertion into the plasmid; (b) insertion into the plasmid and replication in the bacterium.

into a host cell, usually a bacterium, and ensuring its maintenance by incorporating it into a vector which exists and replicates within the host separately from the bulk of the host's genetic material. The control of expression of the foreign DNA can then be studied in the host cell, or large amounts of the specific DNA can be produced for studies *in vitro* such as sequence determination. These specialized techniques are summarized in figure 4.12 and will not be described here in any further detail (see Old and Primrose, 1980; Glover, 1980; Chakrabarty, 1978). The cloned DNA may either be cDNA derived from mRNA, as above, or DNA fragments obtained by digesting chromatin (e.g. with nucleases). The disadvantage of the second procedure is that the correct clone has to be identified (e.g. by hybridization). The availability of large amounts of labelled DNA has been of crucial importance in the study of intervening sequences in DNA (introns) and of multigene families. Research in both these areas depends on the fact that the cDNA can be used as a probe to identify complementary fragments of native DNA derived directly from the cell which hybridize with the cDNA, along with the adjoining DNA sequences (e.g. Nackanishi and colleagues, 1974; Royal and colleagues, 1979).

Split genes—the discovery of introns

An unexpected recent finding is that the base sequence of a structural gene for eukaryotic proteins contains one or more sequences which do not code for amino acids of the protein. All loci so far investigated (e.g. globins, ovalbumin, ovomucoid, immunoglobulins, etc.), with the exception of the histone genes, contain these intervening sequences. These non-coding

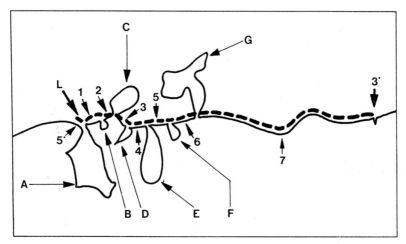

Figure 4.13 Diagram of DNA:RNA hybrids between ovalbumin mRNA and cloned genomic DNA as seen with the electron microscope.
– – – – mRNA; ———— cloned DNA; A–G—introns; 1–7—exons; L—leader sequence. (Redrawn from Gannon and colleagues).

sequences within the loci have been identified by comparing the restriction maps for genomic DNA and messenger-derived cDNA; the genomic DNA coding for the mRNA is found to contain additional restriction enzyme sites. Confirmatory evidence has been obtained by examining hybrids between (cloned) genomic DNA and mRNA in the electron microscope, where the intervening sequences form loops due to the lack of complementary sequences in the mRNA (figure 4.13; Gannon and colleagues, 1979). A terminology has now been established whereby the non-coding intervening sequences are called introns, whilst the coding sequences (and adjoining sequences on either side) are called exons.

Primary transcript RNA which contains the intron sequences can be isolated. It therefore appears that the entire sequence is transcribed into primary transcript RNA, and the intervening sequences are then removed—perhaps in a particular order—by a very specific set of cutting and splicing operations performed by processing enzymes to yield the messenger sequence (figure 4.14; O'Malley and colleagues, 1979).

There is intense speculation as to the role of these non-coding introns within the split gene. One idea, related to the model of Britten and Davidson discussed in chapter 5, is that these sequences could be involved in the regulation of transcription or of the processing of the primary transcript to mRNA. A block in either of these processes would block the formation of the polypeptide chain, or the excised RNA sequences could regulate the expression of other gene action systems. Evidence for this is not available, but could come from studies on such variants as the human thalassaemias where it may be possible to correlate changes in RNA metabolism with changes in intron sequences. Alternatively, it could be that the introns are evolutionary relics. If a polypeptide is derived from the translocation of two (or more) different sequences to the same chromosomal site, the molecule will possess different domains with different recognition sites. In this way allosteric interactions (for example, in end-product inhibition of an enzyme in a metabolic pathway) could evolve (see Brenner, 1967). The intron could then be an inactive linker of the sequences coding for the two domains. This possibility is suggested by studies on the ovalbumin gene family, where the ovalbumin locus has been duplicated twice to give related sequences (X and Y genes). The exon regions of these duplicated loci are well-conserved, whilst the introns are much more variable, both in sequence and in length (Heilig and colleagues, 1980). However, this argument is not rigorous because variability need not necessarily imply lack of importance. Finally, in yeast mitochondria it appears that introns may actually form part of an overlapping locus coding for the RNA processing enzyme. However, the genetic machinery of the mitochondrion is unusual and this finding may well be inapplicable to the eukaryotic nucleus.

Eukaryotic RNA polymerases

In this and succeeding sections, various biochemical studies on the control of transcription are briefly described to illustrate the ways in which gene cloning techniques are advancing our understanding of these processes.

Eukaryotes possess at least three different RNA polymerases. Polymerase I transcribes ribosomal DNA into RNA, polymerase II produces hnRNA which in turn gives rise to mRNA, and polymerase III makes copies of a variety of different low molecular weight RNAs (including the tRNAs, 5S rRNA, and RNAs induced by viral DNA in the cell). The initiation of transcription by polymerase III is the easiest to analyse because the primary transcripts are readily recognized and are not subject to further processing.

Sakonju and colleagues (1980) have identified some of the features of the 5S rRNA sequence in *Xenopus borealis* which are responsible for the initiation of transcription by polymerase III (reviewed by Ford, 1980). Their approach was first to clone a single copy of the 5S rDNA (in a plasmid known as pBR322); this cloned DNA was correctly initiated and terminated both *in vitro* and when injected into a toad oocyte, so it presumably contained all the normal initiation and termination sequences. They then looked at the effect of deleting various lengths of the rDNA sequence by genetic engineering techniques and found that transcription was not affected and a normal transcript was produced when up to about 50 base pairs were removed from the region where it is normally initiated. If extra DNA is inserted 40 base pairs into the sequence, RNA of virtually the correct length is still produced. As termination is unaffected by these procedures, it seems that initiation is controlled by a sequence 50 base pairs into the gene, which initiates transcription about 50 base pairs upstream. The precise initiation point is probably influenced by the sequence in that region which provides the most favourable configuration for the polymerase III.

If deletion is carried out from the other end, deletions up to about the 83rd base pair (as counted from the normal initiation site) do not affect initiation but beyond this point initiation is faulty. It therefore seems that the sequence between bases 50 and 83 of the rDNA is a recognition site for polymerase III initiation 50 base pairs upstream. This site may be common to the different DNAs transcribed by polymerase III; the sequence for tyrosine-tRNA of the moth *Bombyx mori* has a sequence homologous to that of *Xenopus borealis* (bases 55-62 AGCAGGGT) located at a similar site within the gene.

If there is only one RNA polymerase III enzyme, then even if the sequence of the whole initiation site differs between the many loci which are transcribed by it, the relative proportions of the various transcripts cannot be independently varied in different tissues unless there are other factors interacting with the loci and the enzyme. Cloned 5S rDNA from *Xenopus laevis* is not transcribed by purified polymerase III *in vitro*, whereas chromatin prepared from oocytes does produce 5S rRNA with the purified enzyme. A soluble protein, molecular weight 37 000 daltons, has now been isolated from ovarian tissue (Engelke and colleagues, 1980). This produces correct transcription when added to the cloned rDNA/purified polymerase system *in vitro* and binds between the 45th and 96th base pairs of the 55 rDNA sequence. It has no effect on the transcription of cloned methionine-tRNA from *Xenopus laevis*, and therefore has all the properties required of a positive regulatory protein specific for 5S rRNA production.

A model therefore emerges where the initiation of transcription is regulated at a site within the coding sequence interacting with polymerase III and specific proteins. The methods outlined above can be expected to show, in the near future, whether this model is generally applicable to all polymerase III loci, and further the extent to which it applies to transcription of rDNA by polymerase I and the transcription of structural loci to form hnRNA by polymerase II.

Steroid hormones

A rather different approach to the study of the regulation of gene expression has been the biochemical analysis of steroid hormone action in vertebrates. The action of ecdysone in *Drosophila* via a cytoplasmic receptor was mentioned earlier in this chapter; here we discuss progesterone, oestrogen and corticosteroids, whilst genetic studies of androgens are covered in the next chapter.

The induction of egg white proteins, such as ovalbumin in the oviduct of chickens, has been intensively studied by O'Malley and his colleagues (1979) over the last ten years. The induction of these specific proteins, mimicking the natural situation in development of the chick, can be achieved in tissue cultured *in vitro*. Oestrogen is necessary for the maturation of the oviduct, and oestrogen and progesterone then stimulate the production of ovalbumin. It has been demonstrated conclusively that the levels of specific ovalbumin mRNA rise following hormonal induction. Two techniques have been used to show this: transcription of extracted messenger in a cell-free protein-synthesizing system, and hybridization of mRNA with specific cDNA of high radioactivity. However, the rate of messenger formation has not been measured (in contrast to the experiments of Clever and Karlson with ecdysone, where the rate of formation of labelled mRNA was monitored). Thus, although the simplest explanation of these results would be that the hormone stimulation of specific messenger levels is the result of increased transcription, additional or alternative effects on the rate of degradation of specific RNA cannot be ruled out. The data from the cDNA experiment do however indicate that a change in the processing of pre-existing messenger precursor to an active form available for translation is unlikely to be involved, as the cDNA should also hybridize with mRNA precursor.

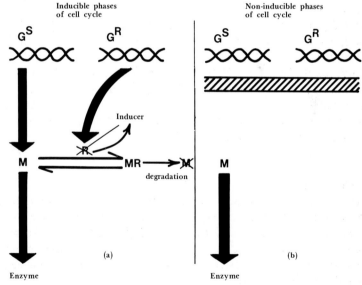

Figure 4.14 A model for the induction of tyrosine aminotransferase by steroids. (a) The structural locus (G^S) produces mRNA (M) coding for the enzyme. In the absence of steroid inducer, M combines reversibly with the regulatory product (R) of the regulatory locus (G^R) which promotes the degradation of M. This leads to a low rate of enzyme synthesis. The presence of inducer inactivates R by an unknown mechanism, leading to high levels of stable M within the cell and hence to an increase in the rate of enzyme synthesis. (b) In the non-inducible phase of the cell-cycle, the rate of enzyme synthesis depends on the level of stable M from the preceding inducible phase. (After Tomkins and colleagues).

In another system, the induction of the enzyme tyrosine amino-transferase in cultured rat hepatoma cells by glucocorticoids, Tomkins' group (Ivarie and colleagues, 1975) have provided evidence from experiments with enucleated cells and with antibiotics inhibiting transcription, that the nucleus may be necessary for the process of de-induction following the withdrawal of steroid stimulation (figure 4.14). This would imply that there is a transcription-dependent effect of the hormone on the rate of specific messenger degradation which might be modulated by steroids. Understanding of the relative roles of transcriptional and post-transcriptional control would be improved if mutant alleles affecting these processes could be identified.

Much attention has also been given to the means whereby steroid hormones bring about such nuclear effects. For many hormones, it has proved possible to isolate cytoplasmic proteins of small amount and high specific affinity for the hormone (putative hormone receptors). The hormone-receptor complexes have been shown to move into the nucleus of target cells in a temperature-dependent step, and to bind to chromatin. Neither the hormones nor the receptor alone will do this and the binding can be shown to be preferential for chromatin which is active in transcription (Scott and Frankel, 1980). In the case of progesterone, the receptor in chick oviduct is a dimer. The B-subunit (molecular weight, 117 000 daltons) binds to non-histone protein DNA complexes but only weakly to pure DNA, whereas the A-subunit binds strongly to DNA but weakly to chromatin. O'Malley and his colleagues (1972) have therefore proposed that the B-subunit specifies the region where binding will occur, and the A-subunit then engages the DNA and facilitates increased transcription; this however remains to be verified.

More recently, cloning techniques have been used to define the DNA sequences which may be responsible for the initiation of transcription. The genes of the ovalbumin family are co-ordinately induced and there are many sequence homologies. Analysis of naturally or artificially varied sequences in such systems may help to identify those important features of the DNA sequence which interacts directly or indirectly through a specific binding with a chromosomal protein, or with the hormone receptor complex (Royal and colleagues, 1979).

In summary, then, there is accumulated evidence that steroid hormones can induce specific protein synthesis. This is probably achieved by the facilitation of transcription of the relevant structural loci, but co-operative effects on RNA processing, mRNA stability and translation may also occur.

Co-ordination of transcriptional control

So far, we have mainly discussed the induction of one specific protein in a target tissue by a hormone. However, the response of the target organ in terms of embryological determination or differentiation will involve the control of many different loci in a defined temporal sequence, specific for each cell-type. One model for this is provided by the study of the puffing of the salivary glands of *Drosophila* during late third-instar larval/prepupal stages described earlier, which can also be mimicked by ecdysone treatment of glands *in vitro*. This system can therefore be manipulated, for example by the removal of ecdysone by washing or by the addition of protein synthesis inhibitors such as cycloheximide. Some of the experiments carried out by Ashburner and his colleagues (1976) may be summarized in a simplified form in figure 4.15, where ecdysone has two actions when complexed with its receptor: to induce an early puff (within 30 minutes) and to inhibit the formation of a later puff. The protein

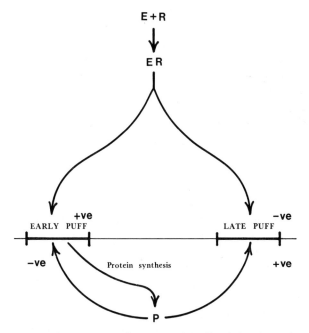

Figure 4.15 A model for the control of the sequential puffing induced by ecdysone (E) and its receptor (R) in *Drosophila melanogaster* salivary gland polytene chromosomes. A product of protein synthesis, P, forms the link between early and late puffs. " + ve" indicates stimulation of puffing; " – ve" denotes inhibition. (After Ashburner).

produced from the first puff will induce a second, late puff when it reaches a sufficiently high concentration to combat the inhibitory effect of ecdysone and will also induce the regression of its own puff in antagonism to the positive effect of ecdysone. In this way, the sequential puffing of the two bands can be understood. It is easy to see how a more complex system involving larger numbers of puffs can be built up from these elements. In principle this system can also be used to investigate the mode of action of juvenile hormone in affecting the competence of the gland to respond to ecydsone in a way appropriate to its stage of development (Richards, 1978). The cloning of the early responsive loci from a variety of tissues (e.g. fat body, imaginal discs, and salivary glands) will establish whether the same coordinating loci are used to induce the later puffs (Ashburner, 1980). This combination of cytogenetic, biochemical and cloning techniques promises to provide unique insights into the coordination of transcriptional activity.

Conclusion

This chapter has presented a brief summary of a number of biochemical approaches to the problem of the regulation of the gene-action system. It should be clear that any step in the chain from chromosome to functional protein may be liable to change. Even at one level—for example, the level of transcription—different mechanisms are possible: supercoiling of the chromosome, binding of inducers to the uncoiled chromosome, regulation of initiation and of polymerase activity, and control of termination. Despite these complexities, the techniques of nucleic acid hybridization, restriction mapping and gene cloning, in combination with other biochemical techniques, have already advanced our knowledge of gene structure and gene regulation. Gene cloning in particular is providing a wealth of information concerning nucleotide sequences and molecular organization of eukaryotic loci, and their functional consequences.

CHAPTER FIVE

GENETIC STUDIES OF REGULATION IN HIGHER ORGANISMS

In prokaryotes, Jacob and Monod's model for the regulation of gene expression of the *lac* operon in *E. coli* was derived from an analysis of the properties of mutant alleles at loci affecting enzyme induction, both singly and in combination in the partial diploid. In particular, the effect of alleles at the regulator locus did not depend on their position (*cis* or *trans*) in relation to the structural loci of the operon (coding for the enzymes β-galactosidase, permease and thiogalactosidase transacetylase), and it was therefore presumed that they exerted their effects by means of a diffusible product. In contrast, mutant alleles at the operator locus were *cis*-dominant and *trans*-recessive, so that their action is confined to the chromosome which carries them. The model suggested by Jacob and Monod on the basis of these observations was subsequently confirmed by biochemical studies which identified the regulator substance and elucidated its mode of action (see Miller and Reznikoff, 1978).

In higher organisms, the gene-action system is more complex and progress in our understanding of regulation has largely come from a combination of the genetic analysis of variant alleles and the study of their effects by appropriate biochemical techniques (some of which were described in the previous chapter). An inherent difficulty is the isolation of suitable variant alleles. In the mouse, it has proved possible to select for biochemical mutants amongst cultured teratocarcinoma cells in very much the same way as auxotrophic mutants can be selected in microorganisms, and such mutant cells have been subsequently incorporated into chimaeric mice (Dewey and colleagues, 1977). Unfortunately, this highly promising approach has yet to show that the mutant carcinoma cells can successfully populate the germ-line and be transmitted by normal genetic means to the offspring of such mice. *Drosophila* chimaeras have also been recovered by transplanting nuclei from *in vitro* cell lines into the posterior region of cleavage stage embryos (Illmensee, 1976). Here again however the donor nuclei are not incorporated into the germ line. The lower eukaryotes such as the fungi and mosses can be grown on defined media and a variety of biochemical mutants selected by conventional techniques. Even here there is the limitation that while one can quickly test for growth rates on different media, it is difficult to devise an automatic screening procedure for identifying characters affecting gene regulation and morphogenesis. Also, since the stage of most interest in the life-cycle is usually the diploid state, a newly-induced mutant allele must be homozygous before it will be detected (as most mutant alleles are recessive, see page 93).

Resort must therefore be made to other means of detecting variation. In experimental organisms, new mutant alleles will from time to time be detected by visual inspection (whether or not a deliberate attempt is made to do this following mutagenesis). In man, the intensive investigation of "sick" individuals, however defined, also results in the identification of inherited disorders. In addition to this, natural populations contain a large amount of genetic variation (which is, itself, a source of controversy among population geneticists). In experimental organisms, this is conveniently reflected in genetic differences between isolates of natural populations. In *Drosophila melanogaster*, chromosomes can be isolated and maintained in specially constructed fly-crosses, and inbred strains of the mouse can be regarded as (non-random) samples of wild type alleles from the same or different populations, depending on the origins of the strains, which are maintained in reproducible form. In this chapter examples of variant alleles identified in these various ways and affecting the expression of other gene loci will be examined in relation to the different levels of regulation outlined in the last chapter.

Genetic studies of heterochromatin

In the last chapter, the idea that chromosome coiling (resulting in the formation of readily staining material known as heterochromatin) is related to gene inactivation was introduced. At this stage, it is useful to consider the notions of *constitutive heterochromatin*, which is always found in the tightly-complexed form (for example, centromeric heterochromatin), and *facultative heterochromatin*, which is variable in its formation; that is, the same region of the chromosome is sometimes heterochromatic, sometimes not. Three examples of facultative heterochromatin will be described; paternal chromosome inactivation in the male mealy bug, variegated-type position effect in *Drosophila*, and X-chromosome inactivation in mammals.

Inheritance in the mealy bug

In the Coccidae, or wingless mealy bugs, genetic studies of morphological characters show that the male expresses and passes on to his offspring only those alleles which he received from his female parent. Cytogenetic analysis shows that both sexes are diploid, but in the male one of the two sets of chromosomes is heterochromatic. This set is discarded in spermatogenesis and must therefore be the paternal set. This means that the lack of expression of paternal alleles is correlated with the formation of facultative heterochromatin. The maternally-derived euchromatic set of chromosomes in the male remains euchromatic when passed to female offspring, but becomes heterochromatic during early embryogenesis in male offspring (figure 5.1). Just how the paternally-derived chromosomes are recognized, become heterochromatic and remain so over many cell generations, is not known. Functionally, this resembles the situation in some Hymenoptera where the female is diploid and the male haploid.

The lack of activity of the heterochromatic chromosomes can also be shown by administering ^{3}H-uridine; autoradiography shows that these

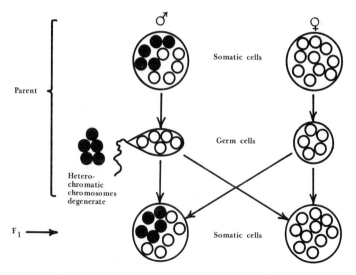

Figure 5.1 Inheritance in the mealy bug. Heterochromatic chromosomes are indicated by solid circles. (Redrawn from Markert and Ursprung).

chromosomes do not incorporate it into RNA as the euchromatic chromosomes do. The heterochromatic chromosomes also incorporate ^{3}H-thymidine later in the mitotic cell cycle. This late replication is presumably also a consequence of the chromosome coiling which renders the DNA inaccessible to DNA polymerase, as it does to RNA polymerase.

In the mealy bug, therefore, facultative heterochromatin is characterized by three properties; deep staining, late replication and, most significantly, lack of genetic activity.

V-type position effect in Drosophila

The effect of heterochromatin on gene expression can be conveniently studied in flies heterozygous for a translocation in which wild type alleles of normally euchromatic genes (for example, the white eye (w) and split (spl) loci, both affecting the compound eye) are brought into close proximity with a block of heterochromatin, usually the constitutive heterochromatin of the centromere or the Y-chromosome (figure 5.2). The heterozygous fly ($w^+\ spl^+/w\ spl$) would normally have wild type (red) eyes but in these specially constructed flies, patches of mutant cells, similar to the clones produced by genetic mosaics (chapter 7) are seen. When the heterochromatin is nearest to the spl^+ allele (figure 5.2), clones of wild type, *split*, and *white split* cells are found, but not *white* alone. This suggests that the heterochromatin is exerting an inhibitory effect on the wild type

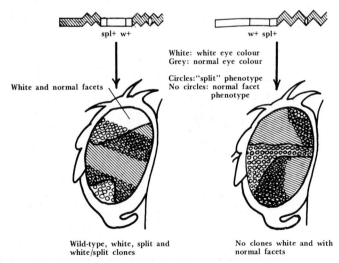

Figure 5.2   V-type position effect in *Drosophila*. (Redrawn from Baker).

alleles of the neighbouring, usually euchromatic, chromosome region by a physical spreading of the heterochromatin into the adjacent region. The mosaic appearance of the eyes will arise then from a variable extent of this spreading in different clones of cells. To test this idea, translocations placing heterochromatin at both ends (figure 5.2) can be examined. The results are consistent with the spreading of heterochromatization (condensation of chromosomal material) along the chromosome. This effect can be observed over a distance of about six map units or about 50 polytene bands.

This is known as a *position effect* because the phenotype depends not just on the combination of alleles present, but on their relative physical arrangement. In this case, it is described as a *variegated* or *V-type* position effect because the effect is not the same in each cell or clone, and hence gives rise to variegation.

X-chromosome inactivation in mammals

Around 1961, Lyon and Russell independently postulated that during the early embryogenesis of female mammals, one of the two X-chromosomes is inactivated at random and becomes heterochromatic. The observation which led up to the "Lyon hypothesis" was that female mice heterozygous for X-linked coat colour alleles are a mosaic of normal and mutant patches (figure 5.3). X-inactivation also accounts for the appearance of a hetero-chromatic structure, known as the Barr body, in most interphase nuclei of cells from female mammals (but not the somatic cells of the mouse) but not in the majority of normal male cells. The Barr body is the cytological

Figure 5.3 Heterozygous $Mo^{vbr}/+$ mouse. Note the pigmented regions (X-chromosome carrying the + allele active) and unpigmented regions (X-chromosome carrying the mutant allele active).

Figure 5.4   Flecked mouse heterozygous for Cattanach's translocation. The melanocytes in the unpigmented regions have the translocated-X carrying the *albino*⁺ allele inactive and are therefore unable to synthesize melanin pigment. Conversely, in the pigmented regions, the translocated-X and hence the *albino*⁺ allele is active. (From McLaren (1976) *Mammalian Chimaeras*, Cambridge University Press; reproduced with permission).

appearance of the heterochromatic X-chromosome. Cattanach (1974) found that autosomal coat colour alleles when translocated to the X-chromosome (translocation T(1 : X)Ct or Cattanach's translocation) gave rise to a V-type position effect (variegation) similar to that described above for *Drosophila*, although their effects were uniform in the normal karyotype (figure 5.4). Such genes, when adjacent to the heterochromatin of the inactive X, are therefore also inactivated by a process analogous to the spreading of heterochromatin seen in *Drosophila*.

In support of the above interpretation is the considerable evidence from the biochemical examination of X-linked electrophoretic enzyme variants, the analysis of RNA production on the chromosome and the time of chromosomal replication. For example, clones from single cultured cells originally derived from individuals heterozygous for an enzyme coded by an X-linked gene such as glucose-6-phosphate dehydrogenase (G6PDH) invariably express one or other electrophoretic form of the enzyme but not both. ³H-uridine is not incorporated into the heterochromatic X-chromosome, and it is late-replicating. These three properties—deep staining, lack of genetic activity as monitored by specific enzyme production and late replication—can be observed concurrently, for example, in crosses between horse and donkey (i.e. female mules and hinnies) where the origin of the chromosomes and the electrophoretic enzyme type can be observed unambiguously (figure 5.5).

When examined in more detail, the "Lyon hypothesis" actually consists of several distinct elements which can each be investigated. These are (i) X-chromosome inactivation is random between maternally- and paternally-derived chromosomes, (ii) X-chromosome inactivation affects the whole of one chromosome, (iii) X-chromosome inactivation is irreversible in the cloned descendants of each cell once it has occurred.

The randomness of inactivation is harder to test than it might appear. Unequal numbers of the two types of clone, apart from statistical fluctuations, might result from cell selection between the two types in the growing embryo. For example, in women heterozygous for the HGPRT⁻ allele (deficiency of hypoxanthine-guanine-phosphoribosyl transferase, leading to the Lesch-Nyhan syndrome of neurological defects in affected individuals), some tissues (e.g. fibroblasts) have normal and deficient clones in equal proportions, whereas others (e.g. erythrocytes, leucocytes) possess only normal clones. This probably reflects the lack of viability of deficient clones in these tissues. However, non-random X-inactivation does seem to occur in some situations. In mules and hinnies, the donkey X-chromosome is more prone to inactivation than the horse's (Giannelli and Hammerton, 1971) but this may only be an artefact of the hybridization between these species. In various species of marsupial (including the euro, wallaroo and red kangaroo) Cooper and his colleagues (1971) have found that the paternally-derived X-chromosome tends to be both inactive and late replicating in most tissues. More recently (West and colleagues, 1977), it has been found that in the mouse, paternal X-inactivation occurs in the early stage of development when embryonic membranes are formed and this gradually becomes random as development proceeds (figure 5.6). Thus, both paternal X-inactivation and random inactivation, at first thought to represent the situation in marsupials and the eutherian mammals respectively, now seem to be the ends of a continuous spectrum.

The usual test of the chromosomal nature of inactivation is to make the *cis* and *trans* double heterozygotes for two X-linked alleles, a and b ($a^+ b^+/ab$ and $a^+ b/ab^+$). In each case, only two types of clone should be found ($a^+ b^+$ and ab, or $a^+ b$ and ab^+). This test was first applied to mouse coat colours by Grüneberg (1969). He compared the expression of the tabby (Ta) mutant (which causes a specific defect in hair structure) and the brindled (Mo^{br}) mutant (which produces a severe reduction in the pigmentation of the coat) in both *cis* and *trans* configuration, and although he found a strong *cis/trans* position effect, he also noted a number of hairs of intermediate phenotype (neither fully mutant nor fully normal). In mice heterozygous for only Mo^{br} ($Mo^{br}/+$), he again noted a less severe expression of the pigmentary defect than is seen in $Mo^{br}/-$ male hemizygotes, and the expression also declined with age. Grüneberg interpreted these data as evidence for incomplete X-inactivation and proposed a complemental-X hypothesis. Recently, the pigmentary deficiency in Mo^{br} mice has been shown to arise from a primary defect in copper homoeostasis (Hunt, 1974); as a result of copper deficiency in the pigment cells, tyrosinase activity in mutant hemizygotes is severely

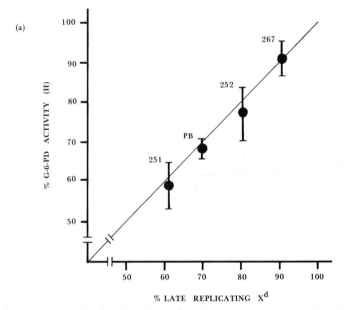

Figure 5.5 X-inactivation in the female mule. The locus for glucose-6-phosphate dehydrogenase (G6PD) is X-linked in both the horse and the donkey, but codes for distinct electrophoretic forms of the enzyme. The donkey X (X^D) is submetacentric, while the horse X (X^H) is metacentric.

(a) Correlation between percentages of cells with late donkey X-chromosome replication and the horse component of G6PD activity in skin fibroblast cultures (251, 252 and 267) and peripheral blood (PB). The line represents the theoretical 1 : 1 correlation. (Redrawn from Cohen and Ratazzi).

(b) Diagram of X-inactivation *in vivo* and the derivation of cell cultures and clones of single cells.

(c) Diagram of electrophoretic patterns of somatic cells and of the two types of clone from single cells. (Modified from John and Lewis).

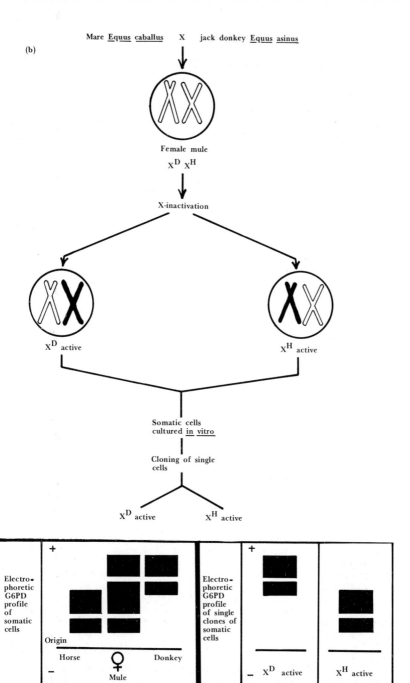

Mare <u>Equus</u> <u>caballus</u>   X   jack donkey <u>Equus</u> <u>asinus</u>

Female mule
X^D X^H

X-inactivation

X^D active X^H active

Somatic cells
cultured <u>in vitro</u>

Cloning of single
cells

X^D active X^H active

(c)

Electro-
phoretic
G6PD
profile
of
somatic
cells

Origin

Horse   ♀   Donkey
              Mule

Electro-
phoretic
G6PD
profile
of single
clones of
somatic
cells

X^D active X^H active

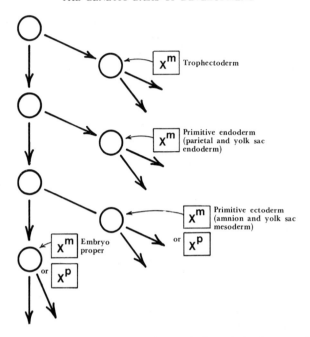

Figure 5.6 Stem-line model for the origin of tissues in the early development of the mouse, as suggested by sequential studies of the inactivation of maternally- or paternally-derived X-linked alleles. X^m denotes an active maternally-derived X-chromosome, X^p an active paternal X-chromosome.

reduced. The copper deficiency is less severe in mutant heterozygotes however, allowing some pigment synthesis in mutant cells and an overall darkening with age. These results are therefore fully consistent with the concept of total inactivation of one or other of the X-chromosomes as proposed by the "Lyon hypothesis". The *cis/trans* test when applied to enzyme determinations either on the hair root follicles of women in a family segregating for the HGPRT and G6PDH alleles, or on clones cultured from single cells of a heterozygote for phosphoglycerate kinase and G6PDH alleles, gives results consistent with total X-inactivation (Goldstein and colleagues, 1971; Gartler and colleagues, 1972). These three loci cover most of the human X-chromosome. Nevertheless, although X-inactivation does seem to be chromosomal in its nature, there is one region which does not seem to be inactivated. The X-linked Xg^a red blood cell antigen is believed to be autonomously determined, yet individuals heterozygous for possession of this antigen (Xg^{a+}/Xg^{a-}) seem to possess the antigen in all cell clones (Fialkow, 1970). The locus for steroid sulphatase is closely linked to Xg^a, and also fails to show dosage compensation (Muller and colleagues, 1980).

As to the reversibility of X-inactivation, the most interesting data come from studies of the germ cells. One hypothesis would be that the X-chromosomes in the female germ-line are never inactivated; this would provide a simple means for ensuring the necessary activity of the maternally derived X in XY offspring. However, it appears that one X-chromosome is in fact inactivated in the female germ cells, only to be reactivated before the start of meiosis. Conversely, in the male germ cells both the X and the Y are heterochromatic throughout meiosis, forming the so-called sex vesicle, but the paternal X-chromosome does become active in the female progeny. Biochemical studies, for example, of HPRT activity in the early embryo, show that both X's are active in early cleavage (Monk and Harper, 1978) even when there is preferential inactivation of the paternal X-chromosome shortly afterwards. However, apart from the particular properties of germ cells which are connected with the changes in X-chromosome activity from one generation to the next, there is little to suggest that the inactivation of an X-chromosome is reversible during normal mitotic division.

Determination of the time of X-inactivation has proved more elusive. Lyon postulated that it should occur during early development to account for the patch size of the coat colour variation. Alternatively, inactivation of the X-chromosome in the retinal pigment cells may occur much later, as was argued by Deol and Whitten (1972) in comparing pigment chimaeras (produced by fusing an albino mutant morula with a non-mutant morula) with mice heterozygous for Cattanach's translocation (with the wild type $albino^+$ gene on the translocated X-chromosome and hence subject to inactivation, and mutant albino alleles on both autosomes). X-inactivation may not necessarily occur simultaneously in all tissues throughout the embryo and the most recent evidence from enzyme activities (Monk and Harper, 1979) is that it may occur at different times as each tissue diverges from a pluripotent stem line (see figure 5.6).

The mechanism of X-inactivation has been the subject of considerable speculation, with very little direct experimental evidence concerning the nature of the controlling factors. It is not even clear whether the change in the X-chromosome is a modification of the DNA (Holliday and Pugh, 1975) or is due to a specific chromosomal protein (Brown and Chandra, 1973). Cattanach (1972), in identifying lines of mice with high or low expression of X-linked mutant loci carried on only one X-chromosome, has

Table 5.1   Proportion of mutant fur in the coat of $Mo^{vbr}/+$ females carrying non-mutant X-chromosomes from various inbred strains (from Cattanach, 1972)

Origin of X-chromosome	Proportion of mutant fur (%)
Strain JU	43.19 ± 1.89
A	50.96 ± 1.88
C57	39.78 ± 2.28
101	39.14 ± 1.75
CBA	51.19 ± 2.12
C3H	56.11 ± 1.51

provided evidence for alleles at a locus on the X-chromosome which probably controls the initial selection of a chromosome for inactivation (Table 5.1), but its mode of action is unknown. Any model has to account for all the observations outlined above, in particular

(a) How is inactivation controlled to leave one active X-chromosome in the cell in all diploid karyotypes (XX, XY, XO with X of either paternal or maternal origin, XXY, XXX, etc.)?
(b) What is the mechanism for choosing the X-chromosome(s) to be inactivated (whether random, or with a bias to the paternal X)?
(c) What is the nature of the somatic memory which renders the inactivation permanent in a clone?

Some models assume that paternal X-inactivation, as in an idealized marsupial, represents the ancestral system, and that random inactivation is a subsequent modification. Unfortunately, this does not greatly simplify the problem of constructing a satisfactory and comprehensive model. There is also an evolutionary aspect to the problem. Mammalian sexual differentiation usually seems to require (a) two active X-chromosomes in the oocyte, (b) the inactivation or elimination of the X in the spermatocyte (see chapter 9 for further evidence), (c) equal X-chromosome activity in the somatic tissues of both sexes, despite the difference in number of X-chromosomes (referred to as dosage compensation). It is not clear what evolutionary factors have led to this situation. Confining the discussion to dosage compensation, one can first observe that aneuploidy for autosomes generally has very severe effects on the organism. As the same is not true for heterozygotes for alleles leading to the complete loss of enzyme activity, which almost always possess 50 % of the enzyme activity of the normal, one possible explanation is that the deleterious effects of aneuploidy are due to regulatory loci of a kind which are more dosage-sensitive. When a Y-chromosome containing linked male-determining factors first evolved, there would be selection for a lack of recombination between it and the X-chromosome. The structural and regulatory loci of the X would then be progressively lost from the Y by mutation, and could not be replaced by back mutation or recombination (a mechanism known in population genetics as "Muller's ratchet"), leading towards effective aneuploidy of one sex compared with the other. There would then be selection for some form of compensation for the aneuploidy. This could take the form of the evolution of regulatory loci on the autosomes instead of the X so that the heterogametic sex would remain viable. An absence of dosage compensation for enzyme structural loci would result, as occurs in butterflies and birds. Alternatively, one X could be regularly lost at mitosis in somatic cells, as occurs in a few mammals (see chapter 9), or there could be transcriptional control of the levels of enzymes coded by the two X-chromosomes, as is found in *Drosophila* (see Lucchesi, 1978). Heterochromatic inactivation of one X, as described in this section, is the final possibility. There are similar alternatives for the regulation of germ-cell sex-chromosome activity. To decide what factors dispose towards one mechanism rather than another, or whether it may be a question of historical accident in each line of evolution, requires a deeper understanding of the developmental mechanisms involved in each system and of the selective forces which would operate in consequence.

The concept of regulatory loci in eukaryotes

As outlined earlier, genetic analysis of mutant alleles can suggest models for regulatory phenomena. The dominance of one allele over another with respect to a particular character might be thought to imply some kind of regulation, or perhaps a threshold phenomenon; epistasis between alleles at different loci could be interpreted as regulation of one locus by the other. To clarify what is meant by the term *regulation*, let us look at these two examples in more detail.

Alleles are empirically described as "dominant" or "recessive" to indicate whether the phenotype of a particular heterozygote resembles one or other homozygote. These terms are purely descriptive. If the heterozygote is intermediate, we say that there is no dominance. There is, however, no such thing as a "dominant allele" in absolute terms. For example, one allele may be dominant to another and recessive to a third (as in allelic series affecting pigmentation in the mouse and *Drosophila*). Alternatively, the description of the dominance relation between two alleles could depend on the particular phenotypic character considered. Thus, the allele for sickle cell haemoglobin is recessive to the wild-type allele for shape of red blood cell and viability (the homozygote has abnormal erythrocytes and dies young), but is dominant for the erythrocyte shape if the cells are examined under reduced oxygen tension (when the erythrocytes of heterozygotes show sickling). With respect to fitness, it is overdominant in areas where malaria is prevalent—Allison (1956) showed that the polymorphism is maintained in these areas largely because the heterozygote is more resistant to malaria than the wild-type homozygote when young, and survives better to reproductive age. However, in the absence of malaria, it is roughly recessive (the heterozygote has about equal fitness with the normal homozygote). At the molecular level, the alleles are codominant (i.e. the heterozygote produces both types of haemoglobin). In this example, the variation in dominance can be understood in terms of the effect of a single amino acid substitution in the α-globin chain of haemoglobin on the function of haemoglobin as the character of the environment is changed. It is also often the case that the dominance relationship of two alleles depends on the genetic background in which they are placed.

Fisher observed that the majority of mutant alleles are recessive to the existing wild-type allele. He suggested that the dominance of the wild-type allele was of selective advantage to the heterozygote, and could therefore have evolved (the "evolution of dominance") as a result of changes in the genetic background. Sewall Wright considered that the dominance of the wild-type allele might be a purely biochemical phenomenon, a physiological consequence of the way in which metabolism proceeds. Thus originated one of the classical arguments in genetics. The mathematical arguments are complicated but the mechanisms involved in the frequent dominance of wild-type alleles continue to be of interest.

Kacser and Burns (1973) have analysed the simplest situation of an unbranched metabolic pathway, controlled by a series of enzymes, in mathematical terms. Let us just consider the simple case of a mutant allele (a) which gives rise to a complete block in the flow of metabolites (flux) along a pathway (synthesizing, say, pigment) in the homozygous state:

$$A \underset{}{\overset{E_1}{\rightleftharpoons}} B \underset{}{\overset{E_2}{\rightleftharpoons}} C \underset{}{\overset{E_3}{\rightleftharpoons}} D \underset{}{\overset{E_4}{\rightleftharpoons}} product$$

Considering the conversion $B \rightleftharpoons C$ and taking the normal homozygote's ($+/+$) enzyme activity (E_2) as 100%, the enzyme activity of the null mutant homozygote (a/a) will be 0%, and it is commonly found that the

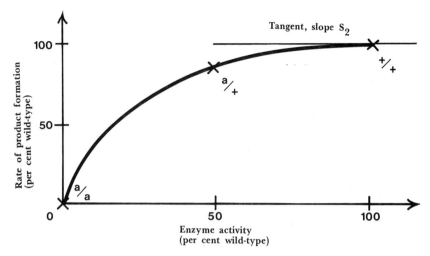

Figure 5.7 Dominance curve.

heterozygote $(+/a)$ has an activity of 50% (i.e. half the normal amount of active enzyme). What can we deduce about the amount of the final product which will be formed?

In the steady state, the rate of production of B from A must be the same as that of C from B, and so on (or the concentration of B would change and the state would not be steady), and must also equal the rate of final product formation. Suppose E_2 were suddenly reduced in amount by 50% then the rate of production of C from B would initially drop by 50% too. However, B would then increase in concentration as it was produced by E_1 from A, and the rate of production of C from B would then increase from its initial drop to 50%. Eventually the pathway would settle down to a new steady state with the final product in general at a level above 50% of the wild-type. A graph of the effect of varying the enzyme activity E_2 on the rate of product formation, which is called a dominance curve, can be drawn (figure 5.7). The precise shape of this curve depends on the relative activities of enzymes E_1, E_3, E_4, etc., compared with E_2, so in general it is not possible to say how much greater than 50% the rate of product formation will be in this heterozygote. Mathematically, it is very difficult to calculate the rate of production. However, the slope of the tangent, S_2, clearly gives some indication of it and is much easier to handle. One can generate, at least in theory, dominance curves for the loci coding for all the enzyme acitivities, one at a time, and these will have slopes S_1 to S_4. It can be shown that:

$$S_1 + S_2 + S_3 + S_4 = 1$$

$\left(\text{or more generally, if there are } N \text{ enzymes in the pathway, } \sum_{n=1}^{n=N} S_n = 1\right).$

Therefore, on average, the slope of a dominance curve will certainly give quite considerably more than 50% of the normal rate of product formation in the heterozygote, which will tend to resemble the wild-type homozygote rather than the mutant homozygote, i.e. wild-type alleles will most usually be dominant with respect to the amount of metabolic product that is formed.

Changing the alleles present at one locus will of course affect the dominance curves of all the other loci, so that dominance relationships at one locus could be affected by selection pressures on the other loci. Real systems are more complex than this hypothetical example; the essential point is however that dominance may be a simple consequence of the properties of the system which is being analysed, and does not necessarily imply any feedback regulation of the rate of enzyme synthesis or breakdown.

Similarly, two loci are said to interact when phenotypes cannot be deduced simply by considering the effects of allelic variation at each of them independently. Such interaction indicates that the effect of one gene-action system is being modulated by another, and this interaction could be due to regulatory mechanisms. Sometimes this interaction is of relatively trivial interest. The simplest case of interaction, or epistasis, can occur when one locus blocks a biosynthetic pathway before the other can act (as when alleles at one locus resulting in an early enzymic block, producing for example a flower without pigment, do not allow alleles at a second locus acting later in the biosynthetic pathway, determining whether the pigment would be red or purple, to be expressed). Although one gene-action system is modulating the expression of another, this occurs after the gene-action systems have produced their functional proteins.

This example of epistatic interaction raises the question of what is meant by the term "regulation". The term can be used in a variety of ways. In the example just given, majority usage would not describe the first locus as regulating the second. If, however, it brought about the same phenotypic effect by an effect on the transcription at the second locus of RNA coding for an enzyme converting red pigment to purple (or *vice versa*), it *would* be described as a regulatory locus. As was discussed earlier, the amount of an active gene product can be affected by actions at any step in the gene-action system so it is arbitrary to restrict the use of the term to any one level, and it is equally arbitrary to regard the active enzyme, rather than (say) the pigment produced by it, as the end of the gene-action system. It is therefore essential to define the level at which regulation is thought to occur, bearing in mind the fact that the term "regulatory locus", when unqualified, carries a strong overtone implying an effect at the level of DNA transcription.

Criteria for discriminating between structural and regulatory loci

Having looked at the theoretical aspects of defining regulatory loci, how are these ideas to be applied in practice? Let us suppose that a mutant allele preventing the appearance of a hormone-induced enzyme has been identified. Residual enzyme activity may often be present, but this does not show the defect to be regulatory because this residual activity could be due to the activity of other, non-mutant, proteins in the enzyme assay system with a limited affinity for the substrate of the deficient enzyme. In the past, this problem has often been approached by trying to show that the putative regulatory allele does not affect the structure of the enzyme (by such criteria as enzyme kinetic properties (K_m, V_{max}), heat stability, electrophoretic mobility and immunological properties) and could therefore be regulatory at the level of DNA transcription, or RNA processing or translation. If a change in one of these properties does occur, then the locus is regarded as the putative structural locus, although it should be realized that post-translational modification of a polypeptide can frequently affect these properties. In the future, with the more widespread use of restriction enzyme mapping using cloned cDNA as a probe, the distinction may become more definitive, as this technique identifies the coding sequence much more directly. The ultimate proof of the identification of a structural locus consists of directly determining the DNA sequence, and showing that it corresponds to the amino-acid sequence of the enzyme, but this will not be feasible for all systems. Alleles at other loci—whether closely linked as promoters or introns, or not—must then be regulatory if they affect the expression of the enzyme. The more traditional criteria outlined above would then be useful in defining the level at which regulation occurs. These different methods of looking at the regulation of the gene action system may be exemplified in studies of the thalassaemias in man and of the regulation of enzyme activity in the mouse and in *Drosophila*.

The primary defects in the thalassaemias

Thalassaemias are a group of inherited disorders in man, defined by unbalanced synthesis of the polypeptide chains of haemoglobin. Different thalassaemias occur in different populations, and they may sometimes be associated with disease resistance (for example, to malaria in the Mediterranean). The chromosomal arrangement of the loci coding for the various globin chains deduced from hybridization and restriction mapping, is indicated diagrammatically in figure 5.8. Hybridization with cDNA indicates that most cases of β-thalassaemia in Asian populations, $\delta\beta$-thalassaemia, and hereditary persistence of foetal haemoglobin can be attributed to deletion of globin loci (Maniatis and colleagues, 1980). The persistence of foetal haemoglobin in the more extensive deletions indicates that in addition to the β and δ genes, regulatory genes that would normally suppress γ-globin production in the adult are also lost. In other types of thalassaemia, the globin genes appear to be intact. For example, in β^0-thalassaemia, although no β-globin is produced, molecular hybridization of a cloned cDNA probe has in most cases failed to detect a deletion or alteration to the β gene. The defect seems to reside in either reduced or absent production of β-globin mRNA (perhaps through a block in RNA transcription or processing), or a specific mutation affecting mRNA translation. In β^+-thalassaemia, a small amount of β-globin is produced, suggesting that the gene is present, and hybridization of reticulocyte RNA shows a corresponding reduction in β-globin mRNA. In the few cases which have been studied, the nuclear RNA of bone marrow erythroid precursor cells contained higher amounts of β-globin mRNA (standardized

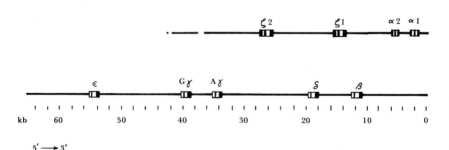

Figure 5.8   Linkage relationships of the human α-like and β-like globin loci. The positions of the adult (α1 and α2) and embryonic (ζ1 and ζ2) α-like loci are shown above; the embryonic (ε), foetal ($^G\gamma$ and $^A\gamma$) and adult (δ and β) β-like loci are below. Non-functional pseudogenes with homologous sequences which are not expressed are present on both chromosomes, but are not shown in this diagram. (Modified from Maniatis and colleagues).

by comparison with the quantity of α-globin mRNA) than the cytoplasmic RNA of bone marrow cells or reticulocytes. These data would appear to indicate a defect in RNA processing or stability.

Other cases of thalassaemia involve structural mutations to the globin loci. An abnormal haemoglobin (Hb Constant Spring) is associated with α-thalassaemia in Asia. The α chains are abnormal, being elongated at the carboxyl terminus by 31 additional amino-acid residues, presumably as a result of a chain termination mutation (UAA for termination to CAA for Gln, the first additional residue). In heterozygous combination with deletion α-thalassaemia alleles, this structural variation is associated with very low levels of the abnormal haemoglobin. Constant Spring α-chains appear to be synthesized at a normal level but degraded more rapidly.

The thalassaemias are a heterogeneous group of disorders and their study has identified genetic variation affecting the globin gene-action system at different levels. Further analysis will add to our understanding of the mechanisms involved. In particular, the sequencing of the abnormal DNA may lead to the identification of the mechanism for the nuclear processing of RNA. A topic of considerable interest is the mechanism of switching from embryonic to foetal to adult globins and it is noteworthy that the genes are linked (figure 5.8) in families (either α-like or β-like) according to their order of expression during development.

The induction of enzymes by androgens in the mouse

The best evidence for a physiological role of the putative steroid hormone receptors of mammals has come from a study of the X-linked mutant allele *Tfm* (testicular feminization) in the mouse. As explained in chapter 9, this allele was discovered because of its effect on the sexual differentiation of XY mice carrying the allele. Studies on target organs of testosterone, such as kidney and submandibular salivary gland have shown that the tissues of these animals are almost completely refractory to the hormone at the cellular level. This lack of response is correlated with the virtually complete loss in most tissues of detectable cytoplasmic receptor protein (Gehring and Tomkins, 1974) for testosterone or dihydrotestosterone (to which testosterone is converted by the membrane-bound enzyme, 5α-reductase, in target cells). Since the receptor is identified solely by its ability to bind labelled testosterone, it is not known whether any mutant form of the receptor protein is present in *Tfm* animals, or if the mutant allele is regulatory for receptor protein. It is also worth noting that the array of proteins induced by testosterone differs between target organs (for example, nerve growth factor is induced in salivary gland, alcohol dehydrogenase in proximal tubule of the kidney). Thus, a common signal (androgen) bound to a common mediator ($+^{Tfm}$ receptor protein) must act on cells which already differ in the way they are programmed to respond; that is, the hormone acts to amplify a pre-existing tissue difference, not to initiate it.

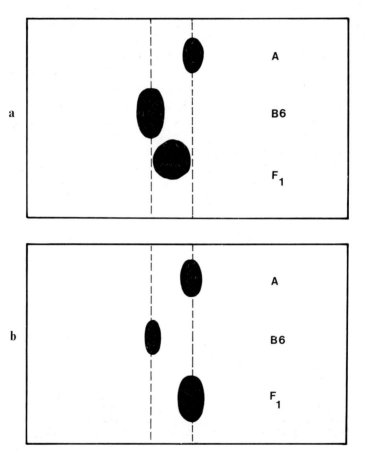

Figure 5.9   (a) Gel electrophoresis of glucuronidase activity in liver or uninduced kidney of strain A, C57BL/6 (B6) and F1 mice. The F1 band is intermediate in position and broader, indicating the formation of heteropolymers from roughly equal numbers of Gus^a and Gus^b subunits.

(b) Electrophoretic pattern of kidney extracts after induction with testosterone. The F1 hybrid enzyme becomes similar to the A parent, indicating the production of an excess of Gus^a subunits. (Redrawn from Ivarie and colleagues).

Paigen and his colleagues (1975) have undertaken extensive investigations of loci affecting the induction and distribution of β-glucuronidase, especially in liver parenchymal cells. Variant alleles were identified in different inbred strains of mice. At one of the loci, three separate alleles have been identified; this is the putative structural locus, known as *Gus*, because the *Gus*[a] and the *Gus*[h] alleles differ from *Gus*[b] by electrophoretic mobility (figure 5.9) and thermal stability respectively. The enzyme specified by this locus on chromosome 5 is a tetramer of identical chains (molecular weight 70–75 000 daltons). β-glucuronidase activity is found both in the lysosomes and the endoplasmic reticulum. As these alleles affect the two sites, the same enzyme is present at both. In the backcross generation between strains of mice carrying the *Gus*[a] and the *Gus*[h] alleles, the type of enzyme present also affects its intracellular distribution between these two sites. This suggests that the structural difference between *Gus*[a] and *Gus*[h] proteins lies within a region affecting their recognition by the intracellular transport system (or perhaps that there is a closely linked gene affecting distribution; but this is a more complicated model).

The lack of microsomal but not lysosomal β-glucuronidase activity in

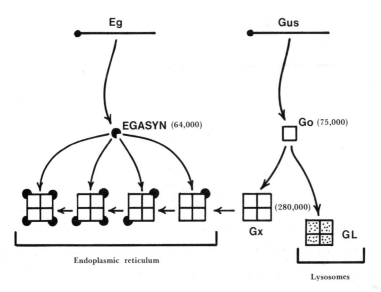

Figure 5.10 Intracellular distribution of β-glucuronidase. The active enzyme (G_x or G_L) is a tetramer of G_0. Attachment of the active enzyme to the endoplasmic reticulum is achieved by the binding of egasyn molecules. (Redrawn from Ivarie and colleagues).

the YBR inbred strain led to the identification of another gene affecting the cellular distribution of this enzyme, the *Eg* gene on chromosome 9. This gene codes for a protein termed egasyn, molecular weight 64 000, which can bind with one form of the tetramer specified by the *Gus* locus and enable it to be bound to the endoplasmic reticulum (up to 4 molecules of egasyn per tetramer—figure 5.10). The Gus^h protein presumably differs from Gus^a in the binding site for egasyn. The recessive Eg^0 mutant allele in the YBR strain does not affect the other proteins of the endoplasmic reticulum; it is specific to β-glucuronidase.

In addition to the above characteristics, β-glucuronidase is induced in kidney by testosterone, and Paigen and associates have identified genetic factors that regulate this induction. Strains C57BL/6 and A differ in the inducibility of the enzyme, C57BL/6 responding much more slowly and achieving a lower level of enzyme activity (figure 5.11). This effect is specific to β-glucuronidase (as alcohol dehydrogenase and arginase are induced similarly in the two strains, and the response of hypertrophy by the proximal tubule cells is the same in both). The effect is also limited to the kidney. F_1 animals are intermediate in inducibility, and the F_2 segregation into 3 discrete classes is consistent with a single gene control of

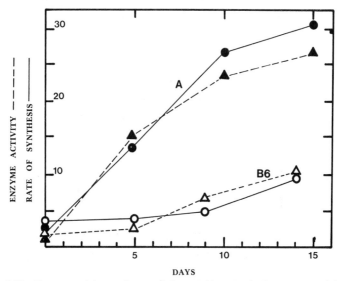

Figure 5.11 Enzyme activity, and rate of glucuronidase synthesis as measured by pulse labelling, in the kidneys of A and C57BL/6 (B6) mice following injections of testosterone at 3 day intervals. (Redrawn from Ivarie and colleagues).

inducibility. No authenticated recombinants between this locus regulating inducibility (*Gur*) and the structural locus (*Gus*) have been obtained. Pulse-labelling studies with radioactive leucine have shown that the increases in enzyme activity and in the number of enzyme molecules (as determined immunologically) arising from the possession of the high inducibility Gur^a rather than the low inducibility Gur^b allele, are due to an increased rate of enzyme synthesis. In the double heterozygote $\underline{Gur^a\ Gus^a}/\underline{Gur^b\ Gus^b}$, an excess of Gus^a over Gus^b enzyme molecules is made (figure 5.9), indicating that the *Gur* locus does not act through a diffusible product affecting both chromosomes, but is *cis* acting (i.e. regulating the *Gus* locus that it is physically adjacent to on the chromosome).

A further type of variation has been found, where the developmental pattern with age of β-glucurodinase activity in the liver (and also in other tissues) differs significantly between strains C3H and DBA (figure 5.12). A

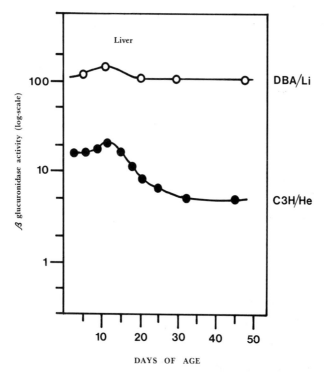

Figure 5.12 Variation in β-glucuronidase activity during liver development in two strains of mice. (Redrawn from Ivarie and colleagues).

Table 5.2 The use of recombinant-inbred strains in identifying additional loci controlling induced kidney levels of β-glucuronidase (data from Watson and Paigen, 1978)

Strain	Electrophoretic type (Gus allele)	Induced enzyme activity	Rate of synthesis K_S	Rate of loss K_L	Interpretation
C57BL/6(B)	b	B (low)	B (low)	B (low)	
BABL/c(C)	a	C (high)	C (high)	C (high)	
CXB D	a	C	C	C	Non-recombinant for genes controlling K_S and K_L
CXB J	b	B	B	B	
CXB K	b	B	B	B	
CXB G	a	New high	C	B	Recombinant for genes controlling K_S and K_L
CXB H	a	New high	C	B	
CXB E	b	C	B	New low	Identifies at least 2 loci controlling K_L: the parental strains differ at both loci and this strain is recombinant for the 2 loci

simple backcross between these lines, and the examination of different (congenic) strains of C57BL/6 mice carrying the *Gus* chromosomal region of other inbred strains, both show that there is a single locus (*Gut*) closely linked to *Gus* (no recombinants have been found) which controls the difference in the temporal pattern of enzyme activity between strains. Like *Gur*, it affects the rate of enzyme synthesis rather than degradation, but unlike *Gur*, each allele of *Gut* affects the activity of both *Gus* structural genes (i.e. it acts *trans*).

In addition to these important effects on β-glucuronidase activity initially identified in different strains, some account should be taken of additional genetic variation that may modulate the activity of this enzyme, yet not give rise to distinctive phenotypes in crosses between strains, either because the differences are too small to fall into non-overlapping classes or because they are dependent on multiple gene loci. The study of recombinant-inbred strains offers an approach to the analysis of characters of this type. Recombinant-inbred strains are produced by the intercross of two inbred strains and the subsequent establishment of a number of inbred lines from the F_2 progeny. As a result, genetic differences between the two original strains are initially reassorted and become "fixed" in different combinations in the different lines. In many cases, the genetic basis for relatively small differences in enzyme activity can now be determined. Watson and Paigen (1978), by utilizing the recombinant-inbred strains between the C57BL/6 and BALB/c strains, were able to identify at least two further genes affecting the rate of loss of kidney β-glucuronidase (table 5.2). Other enzyme activity variants have been analysed in a similar way.

In summary, the *Gus* locus is the structural gene for β-glucuronidase. The unlinked *Eg* locus affects the processing of the enzyme and gives intracellular location, whilst at least two other loci (identified by recombinant-inbred strain analysis) control its degradation and secretion out of the kidney proximal tubule cell. The temporal and tissue distribution is affected by the *Gut* locus (closely linked or identical to *Gus*), whilst the response to testosterone is mediated by the *Tfm* locus (controlling the receptor for the steroid but not kidney-specific) and the kidney-specific *Gur* regulatory locus (closely linked or identical to *Gus*). The detailed biochemical mechanisms remain to be elucidated: in particular, it is not clear that *Gur* and *Gut* are distinct from *Gus*, and they could map within it (e.g. within an intron).

This system is the most completely described in mammals at present, but similar closely linked regulatory and structural loci have been identified for δ-amino-laevulinate dehydratase (Coleman, 1971) and β-galactosidase in the mouse (Paigen and colleagues, 1976). In other cases, such as for α-galactosidase (Lusis and West, 1978) and catalase (Ganschow and Schimke, 1969), unlinked regulatory loci have been identified which, in the latter case, act by specifically affecting the rate of degradation of the enzyme protein. A similar degradative control of the activity of the adrenal catecholamine-synthesizing enzyme, phenylethanolamine N-methyltransferase, was identified by Ciaranello and Axelrod (1973) who showed that the resulting different levels of catecholamines were correlated with differences in aggressive behaviour of the mice (Ciaranello, 1978).

Regulatory loci in *Drosophila*

The analysis of strains of *Drosophila melanogaster* carrying variant activity alleles for aldehyde oxidase has led to the identification of a closely-linked *cis*-acting locus controlling the level of enzyme synthesis in the pupa (Dickinson, 1975; figure 5.13). A low apparent frequency of recombination (about 2%) between this temporal locus and the structural gene was

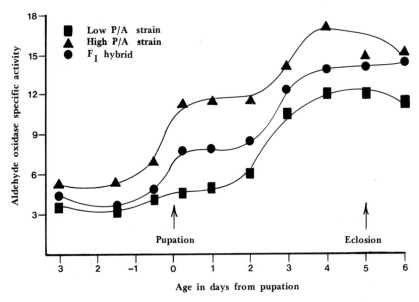

Figure 5.13   Aldehyde oxidase specific activity (units/mg protein) as a function of developmental stage in a low P/A (ratio of pupal to adult activity) strain, a high P/A strain and their F_1 hybrid. (Redrawn from Dickinson).

obtained. However, since only one recombinant class was observed and this was the type that could have arisen by errors in determining enzyme activity, the linkage is probably much tighter than this. Variant alleles affecting the activity of amylase have also been reported (Abraham and Doane, 1978), although in this case the regulatory gene controls the tissue distribution of the enzyme during development; young adult flies homozygous for the *map*^{b} or *map*^{c} alleles respectively either have reduced levels of, or lack, amylase activity from the posterior midgut compared with flies homozygous for *map*^{a}. After 2–3 weeks, however, the *map*^{b} and *map*^{c} flies come to resemble *map*^{a} with activity throughout the midgut. The *map* regulatory gene is an authenticated 2 map units from the structural gene and acts *trans* in determining posterior mid-gut amylase activity.

Extensive biochemical and genetic fine structure analysis of the structural gene for xanthine dehydrogenase in *Drosophila* (the so-called rosy, *ry*, gene after its phenotypic effect on pigmentation in the eye) by Chovnick and associates has enabled closely-linked regulatory loci to be precisely mapped (Chovnik and colleagues, 1976; McCarron and colleagues, 1979). The recovery of a high (i409) and a low (i1005) activity variant, both located on the centromere side of the structural gene mutant, identified a *cis*-acting regulatory gene. By extrapolating from the known map unit size of the structural gene and assuming that the XDH polypeptide of 150 000 daltons contains 1363 average size (110 daltons) amino acids, the size of the regulatory gene is estimated as between 2.8–4.4 kilobases (figure 5.14). This is, therefore, the only system in higher

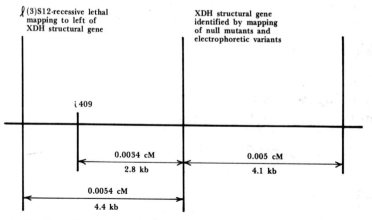

Figure 5.14 Map of *rosy* region, showing size estimates of structural and regulatory loci. (From Chovnick and colleagues).

organisms where it has been possible to estimate the size of a *cis*-acting regulatory locus.

A model for gene regulation in higher organisms

Although there are certain similarities between the bacterial operator and the *cis*-acting regulatory genes identified in higher eukaryotes, Paigen has been careful to stress in the case of β-glucurodinase that "the properties of *Gur* variants which differ in the duration of the lag period, rate of induction and final rate of enzyme synthesis, but have constant basal levels, are unlike those of any known microbial variants, suggesting that the processes of regulation in higher eukaryotes may be quite different from those in micro-organisms". Similar caution should be taken in interpreting the other examples of gene regulation as parallels of microbial systems. Unlike prokaryotes, the higher eukaryotes requires a complex and sensitive system of gene regulation so that precise levels of enzyme activity may be maintained (and the levels may differ in different tissues), and different sets of genes may be activated in different tissues or in the same tissue at different developmental stages. This would seem to require a mechanism for the selection of different yet overlapping sets of genes for activation, either by a single stimulus such as a hormone acting as a trigger, or as a result of the changing physiological state of the cell.

In 1969, Davidson and Britten proposed such a model (subsequently revised in 1973) of gene regulation in higher organisms in which individual structural genes are controlled by adjacent regulatory sequences, but co-ordinate gene activity is achieved by activator proteins coded by integrator gene sets acting within the nucleus. A further feature of the model is that each structural gene is preceded by a number of these regulatory sequences, each responsive to a different activator protein. The integrator gene sets are, in turn, controlled by a sensor gene that is sensitive to specific external stimuli such as a hormone or other information molecule entering the nucleus (figure 5.15). Depending therefore on the nature of the external stimulus, different integrator gene sets would be activated, leading to the production of a particular group of activator proteins and hence the co-ordinate activation of a number of gene loci.

The closely-linked *cis*-acting regulatory loci described previously are clearly analogous to the adjacent regulatory gene sequences proposed by Davidson and Britten, and the *trans*-acting elements, whether linked as in the case of *Gut* or separate from the structural gene, as in the case of *map* or the α-galactosidase regulator, may also identify analogous genes that act

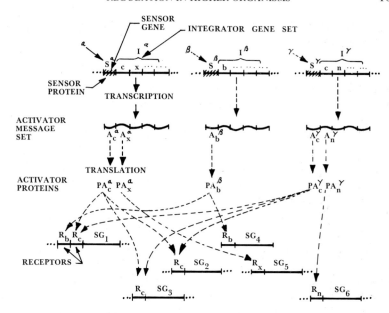

Figure 5.15 Interrelationships in the Britten-Davidson model of the regulatory system in eukaryotes. α, β, γ are effectors (external signals producing pleiotropic responses) S^α, S^β, S^γ are sensor regions in the genome responding to α, β, γ, consisting of RNA and sequence-specific proteins which bind the appropriate effector. I^α, I^β, I^γ are integrator gene sets, coding for combinations of activator proteins c, x..., b..., c, n,...., A_c^z, A_x^d, A_b^β, A_c^γ, A_n^γ are the activator messenger RNAs derived from these integrator gene sets: PA_c^z, PA_x^z, PA_b^β, PA_c^γ, PA_n^γ, are the activator proteins translated from the activator mRNAs R_b, R_c, R_x, R_n are receptor sequences recognized by the corresponding activator proteins. SG_1, SG_2, SG_3, SG_4, SG_5, SG_6 are structural genes each linked to a particular receptor sequence(s) and transcribed when these are activated by the appropriate activator protein. The structural genes in this diagram therefore belong to four batteries, each of which will be co-ordinately induced (b — SG_1, SG_4; c — SG_1, SG_2, SG_3; x — SG_5; n — SG_6). Effector α will induce batteries c and x; β will induce b; γ will induce c and n, as seen by following the dashed lines. (After Britten and Davidson).

via a diffusible product. Alternatively, the *trans*-acting regulatory loci may represent a further level of regulation between the *cis*-acting transcriptional control and the integrator gene sets. Whether the Tfm^+ gene can be accommodated within this scheme is less certain; if Tfm^+ codes for or controls the production of the hormone-receptor protein, then this gene may also represent an additional step in information processing in the cell. Many of the regulatory loci so far identified in higher organisms are consistent with this model and a number of additional predictions follow from it. Mutations in integrator gene sets may be expected to simultaneously affect a number of related gene loci. An example of such a mutation was found by Arst (1976) in *Aspergillus nidulans*; the *int A* gene regulates the activity of three separate structural gene loci coding for acetamidase, GABA transaminase and GABA permease. Alternatively, in other cases the effect of integrator gene mutation may be seen as pleiotropic gene action (see chapter 8 for a fuller discussion of this phenomenon). The reduction in the rate of synthesis of a number of plasma proteins (albumin, α-fetoprotein, transferrin) accompanying the deletion of the *albino* gene in the mouse (Garland and colleagues, 1976) suggests that a regulator gene of this type has been lost along with the *albino* locus. Mutations in sensor genes would likewise give rise to pleiotropic effects. In many cases, mutations in either sensor genes or in integrator genes may so disrupt normal cellular homeostasis as to result in cell lethality. Many such mutations will therefore not be recoverable by conventional means.

The Davidson and Britten model is consistent with a number of observations on the nature of gene regulation in higher organisms. This is not of course proof of its correctness, since other interpretations are possible, and it is almost certainly an over-simplification of the true situation within the cell. Also, the model does not offer a mechanism for the control of a number of the developmental processes discussed previously, especially the rearrangement of the immunoglobulin V genes during lymphocyte maturation and the other changes in genomic DNA that accompany development in certain species, and does not take account of the regulatory mechanisms which act after the transcription of DNA to hnRNA. Many other interactions between different gene loci have been described and are the topic of the next chapter.

CHAPTER SIX

GENE INTERACTIONS IN DEVELOPMENT

In the last chapter we considered factors affecting the expression of one gene-action pathway within the cell. In many cases, however, the interacting gene loci have been less precisely defined and even the demonstration of their presence may depend on special experimental conditions.

In a classical series of experiments, Waddington (1953) subjected outbred *Drosophila* larvae to a severe environmental shock. This caused a characteristic interruption in a wing vein (a phenocopy of the mutant gene *crossveinless*). Breeding from the affected flies followed by phenocopy induction in their progeny over successive generations eventually led to the establishment of a stock of flies that developed abnormal wings in the absence of the environmental shock. This apparent Lamarckian inheritance of an acquired character, known as genetic assimilation, can be shown to arise from the selection during the intervening generations of a combination of alleles at several loci which together produce the abnormal wing (Falconer, 1960). This is most easily understood in terms of a threshold model, where individuals with a level of some morphogenetic determinant below (or above) a certain critical level develop the abnormal phenotype (figure 6.1). This level can be produced either by the original mutant alleles or by selecting individuals near the threshold as revealed by the environmental shock, and breeding from them. If selection is repeated in successive generations, the level of morphogenetic determinant will progressively approach the threshold until flies are produced that express the developmental defect in the absence of the environmental shock.

This example demonstrates that a developmental process depends on the action of a combination of alleles at many loci, and the effect of a mutant gene is determined not just by itself, but by the way it combines developmentally with alleles at other loci. Developmental stability (or

111

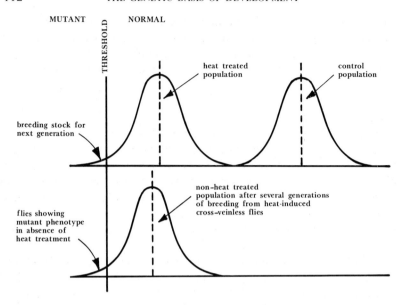

Figure 6.1 Distribution of hypothetical morphogen controlling the appearance of the *cross-veinless* phenotype in control and heat-treated *Drosophila* pupae.

canalization) depends on the ability of the organism to withstand environmental shock, as reflected in the ease with which the level of morphogenetic determinant can be reduced to the critical threshold level.

In a similar experiment, Rendel (1959) attempted to select directly for a change in the number of large bristles on the scutellum of the *Drosophila* fly. In normal stocks, this is a highly canalized character and the number is invariably four, so selection is impossible. However, in the presence of the *scute* allele, development is destabilized (the level of hypothetical morphogen is reduced below the critical level) and outbred stocks show a variable reduction in the number of these bristles; selection can now proceed. When the *scute* allele is removed by outcrossing after several generations of selection, the number of bristles is found to be no longer four, but increased or decreased depending on the direction of selection. The action of *scute* (or other major mutation) is therefore to reduce canalization, with the result that many loci affecting this process are now exposed to selection.

There are many other examples from *Drosophila* and other species of the interaction of gene loci in the development of a particular character or in the expression of a particular mutant gene. Gene interactions can be studied in many other developmental systems. Notable examples that will be discussed are the position effects within complex loci, the use of somatic cell hybrids and the analysis of controlling elements (particularly in maize).

Complex loci in *Drosophila*

The *bithorax* series of pseudo-alleles has been extensively studied by Lewis (1963). They are of particular interest developmentally since they belong to the group of homoeotic mutants that result in a change from one differentiative capacity to another. In the case of *bithorax*, the different alleles affect the determination of structures appropriate to the particular segment of the adult fly. (Note that some alleles are dominant to wild type, which affects the interpretation of the complementation tests). Table 6.1 describes the effects of the mutant alleles.

Normally, the mesothorax bears the wings and the second pair of legs, and the metathorax bears the halteres—as in other Diptera—and the third pair of legs, whereas the first abdominal segment has no appendages. The

Table 6.1 The effects of some mutant alleles of the *bithorax* series

	Type of deviation from wild phenotype			
Genotype	A	B	C	D
bx^3 (bithorax) bx^3	+ + +	0	0	0
Cbx (contrabithorax) +	0	0	0	+ + + +
Ubx (ultrabithorax) +	+	0	0	0
bxd (bithoraxoid) *bxd*	0	0	+ + +	0
pbx (postbithorax) *pbx*	0	+ + +	0	0

Note: *Ubx* is lethal (at larval stage) when homozygous.
    + indicates degree of deviation from wild phenotype.
Type A—anterior metathorax resembles anterior mesothorax.
    B—posterior metathorax resembles posterior mesothorax.
    C—posterior metathorax resembles posterior mesothorax and first abdominal segment resembles anterior metathorax.
    D—posterior mesothorax resembles posterior metathorax.

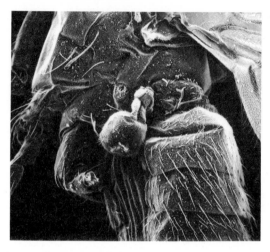

Figure 6.2 *bx/bx* fly, with enlarged halteres.

different alleles of the *bithorax* series can severely disrupt this pattern (figure 6.2). For example, *bx pbx/bx pbx* flies will have an entire meta-thorax resembling the mesothorax, producing a four-winged dipteran, and *bxd/bxd* animals are eight-legged insects. The genetic map is shown in figure 6.3. The complex of pseudo-alleles covers two bands on the cytogenetic map of the salivary gland chromosomes.

The phenotype of the animals in complementation tests is given in table 6.2. When these are examined alongside the effects of the individual alleles shown in table 6.1, several interesting features emerge. For example, the phenotypes of $bx^3 Cbx/+ +$ and $Cbx Ubx/+ +$ are actually less severe for the type D effect than those of $+ Cbx/+ +$ and $Cbx +/+ +$ respectively, so that some suppression of mutant effects takes place in *cis* combinations. In the *trans* combinations, complete complementation occurs between some alleles (e.g. bx^3 and *bxd*, *Cbx* and *bxd*, bx^3 and *pbx*) but not between others (e.g. *bxd* and *pbx*, bx^3 and *Ubx*, *Ubx* and *bxd*, *Ubx* and *pbx*, *Cbx* and *Ubx*). Furthermore, in those combinations between recessive alleles where complementation does not occur, there is a

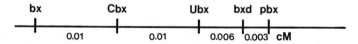

Figure 6.3 Genetic map of the *bithorax* region in *Drosophila*.

Table 6.2 Summary of the phenotypes of double heterozygotes for mutant alleles in the *bithorax* series.

Mutant double heterozygote	*Deviation from wild phenotype*			
	A	B	C	D
bx³ and *bxd* cis	0	0	0	0
trans	0	0	0	0
bx³ and *pbx* cis	0	0	0	0
trans	0	0	0	0
bxd and *pbx* cis	0	0	0	0
trans	0	+ + +	0	0
bx³ and *Ubx* cis	+	0	0	0
trans	+ + +	+	0	0
Ubx and *bxd* cis	+	0	0	0
trans	+	+ + +	+ + +	0
Ubx and *pbx* cis	+	0	0	0
trans	+	+ + +	0	0
bx³ and *Cbx* cis	0	0	0	+ +
trans	0 to +	0	0	+ + + +
Cbx and *bxd* cis	0	0	0	+ + + +
trans	0	0	0	+ + + +
Cbx and *pbx* cis	0	0	0	+ + + +
trans	0	0 to +	0	+ + + +
Cbx and *Ubx* cis	+	0	0	+
trans	+ +	+	0	+ + +

tendency for the type of effect which is seen to be characteristic of the right-hand allele of the pair (rather than the left one) if they affect different parts of the insect. Thus, $bxd +/+ pbx$ individuals resemble *pbx/pbx* flies rather than *bxd/bxd* flies. In combinations which are homozygous pbx^+/pbx^+ and complementation does not occur, the type B effect characteristic of *pbx* is seen despite the lack of mutant alleles at this locus. In other words, there is a polarity effect in the complementation of these closely-linked mutants. One simple model to account for this would be to suppose that normal development requires the appropriate synthesis of each wild type gene product, and that selection of the appropriate locus for transcription depends on the movement of some factor along each chromosome (perhaps as a result of local effects within the cells of the different thoracic imaginal discs). In the *trans* double heterozygote $ab^+/a^+ b$, the ab^+ chromosome would be blocked at the first locus; however, the wild type allele on the $a^+ b$ chromosome could be appropriately expressed. For the second locus, neither chromosome could be expressed, ab^+ because it has already been blocked, $a^+ b$ because it carried the mutant allele b.

The model also explains why $Ubx + / + bxd$, for example, show type B as well as type A and C effects. Since both chromosomes carry mutants to the left of pbx^+, transcription of this locus will necessarily be blocked. Interactions between one chromosome and the other, as well as between loci on the same chromosome, may be involved, as indicated by the effects of translocations (indicated by R followed by the rearranged alleles); $R(bx^{34e} +)/+ Ubx$ has a more extreme phenotype than $bx^{34e} + / + Ubx$, but $R(bx^{34e} +)/R(+ Ubx)$ does not. Likewise, $R(bx^+ Cbx^+ Ubx^+)/bx^+ Cbx^+ Ubx^+$ has an extreme type B and type C phenotype even though no mutant alleles are present. Somatic pairing of chromosomes is known to occur in *Drosophila*, and could provide the basis of an explanation for these abnormal phenotypes (as well as the complementation which is observed in some combinations) if the factor responsible for selecting loci to be transcribed can cross from one chromosome to the other at certain sites. A detailed model accounting for these observations is difficult to devise and test experimentally but the conclusion that the stable position effects (S-type position effect) between these alleles are important in the coordination of development seems inescapable. Further aspects of the action of homoeotic mutants, and the ways in which they may bring about their effects, are discussed in chapter 7.

Complex loci without homoeotic effects are also known in *Drosophila*. For example, the *miniature* (wing) mutants, the *lozenge* (eye shape) mutants and the *rudimentary* (wing) mutants each form a series of closely-

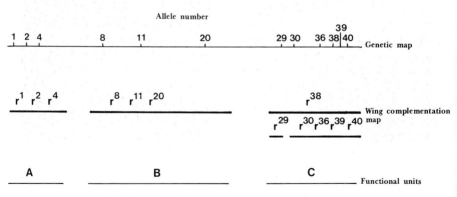

Figure 6.4 Genetic and complementation map for a number of *r* mutants in *Drosophila*. The functional units A, B and C correspond to the enzyme activities dihydroorotase, carbamyl phosphate synthetase and aspartate transcarbamylase respectively.

linked alleles which mainly fail to complement in *trans* heterozygotes. The *rudimentary* (*r*) series is especially interesting since the biochemical basis for their aberrant behaviour is largely understood (Rawls and Fristrom, 1975). The complementation map is essentially co-linear with the genetic map (figure 6.4) and identifies the genes coding for the first three enzymes (carbamyl phosphate synthetase, aspartate transcarbamylase, dihydro-orotase) in the pyrimidine biosynthetic pathway. These three enzyme activities probably exist in a single multi-enzyme complex (Brothers and

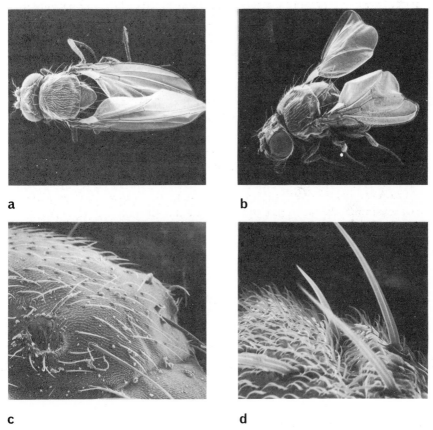

a   b

c   d

Figure 6.5 The *dumpy* mutant in *Drosophila*.
   (a) Wild type;
   (b) *dp/dp* mutant showing oblique wings;
   (c) pit or vortex in posterior thorax;
   (d) comma-shaped depression in anterior thorax.

colleagues, 1978), and the exceptional complementation behaviour of certain mutants may arise either from protein–protein interactions between the products of the three genes in the formation of this complex, or from inter-allelic complementation between certain alleles in the production of individual enzyme activities. The phenotypic effects of reduced wing size, larval inviability and maternally-inherited embryonic lethality may arise from a general depression in nucleic acid (RNA and DNA) synthesis in mutant cells and tissues consequent on the reduced level of pyrimidine synthesis *in vivo*.

The dumpy (*dp*) series of pseudoalleles provide a further level of complexity. In addition to complex complementation patterns, four different phenotypic effects (figure 6.5) are associated with this locus (Carlson, 1959). These are oblique (*o*—truncated wing), vortex (*v*—whorls or pits on the posterior or occasionally anterior thorax), comma (*c*—depressions on the extreme anterior thorax), and lethality (embryonic and larval). Different alleles may express more than one of these defects, and heterozygotes for different alleles will express only those defects that they share in common (e.g. dp^o/dp^{ov} will express *o* only). In each case, morphological effects arise from abnormal chitinous structures, and the lethality can also be attributed to malformed chitinous tracheae (Metcalfe, 1971). A recent study by Blass and Hunt (1980, 1981) of the synthesis of chitin precursors in *dp* larvae also suggests a primary role of chitin synthesis in the expression of the *dp* defects; the effects of the different alleles may depend therefore on the differential expression of a relatively simple lesion in chitin production in the different larval and adult tissues.

In the examples described above, genes of related function are organized into single units. In the case of the *bx* series, this could reflect the need for sequential gene action during development. Alternatively, the interactions between the different *r* alleles probably arise from their involvement in the production of a multi-enzyme complex; the functional unit may exist to ensure co-ordinate expression of the component genes. The possibility exists therefore that the aberrant interactions of alleles at other complex loci also arise from the the production of multi-enzyme complexes.

Mammalian opioid peptides

Other examples of the organization of genes into functional units are the histone and rRNA gene clusters (page 16); in both cases, this organization may enable co-ordinate expression and is most probably a result of evolution by gene duplication. A similar interpretation of the organization

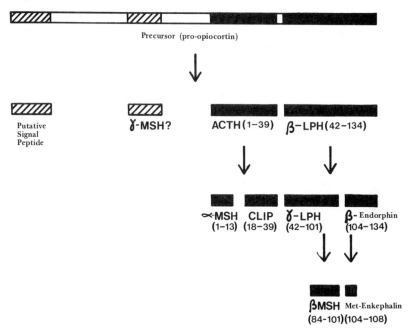

Figure 6.6 Derivation of pituitary peptides from a common precursor, as deduced from the nucleotide sequence of cloned bovine cDNA.

ACTH—adrenal corticotropin; β-LPH—β-lipotropin; α-MSH—α-melanotropin; CLIP—corticotropin-like intermediate-lobe peptide; γ-LPH—γ-lipotropin; β-MSH—β-melanotropin; Met-enkephalin—methionine enkephalin; γ-MSH—putative γ-melanotropin. (After Nackanishi and colleagues).

of the genes coding for the mammalian opioid peptides is warranted. The analysis of a cloned cDNA prepared from mRNA for bovine pro-opiocortin (the polypeptide precursor of adrenocorticotropin, ACTH and β-lipotropin) has shown that this mRNA also codes for a sequence homologous to α- and β-MSH (termed γ-MSH) with unknown function (Nackanishi and colleagues, 1979). Pairs of basic residues (arginine and lysine) between these sequences are sites for the proteolytic degradation of the precursor to adrenocorticotropin (ACTH), β-lipotropin (β-LPH) and β-endorphin: α-LPH is further broken down to form β-MSH (figure 6.6). Other peptides may also be formed as these molecules are degraded *in vivo*. β-endorphin, β-LPH, ACTH, and β-MSH are found in neurones in the same regions of the brain and may be co-ordinately expressed; there is independent biochemical evidence for their formation in the brain from a common precursor. Methionine-enkephalin, which could be formed from

β-endorphin, has a different distribution in the brain and is also synthesized locally in the gut; it is possible therefore that it is formed from a separate (but related) locus which could also produce yet further peptides. A fourth region with sequence homology to MSH is found within the precursor sequence which is apparently not found as a separate molecule. It therefore appears that pro-opiocortin has evolved by repeated gene duplication, followed by base substitution, deletion and insertion in the various regions of the locus. A wide variety of related peptides are then produced by post-translational modification. In the foetal pituitary of the rhesus monkey, β-MSH, CLIP (corticotropin-like intermediate lobe peptide) and β-endorphin predominate, whereas in the adult there is more ACTH, α-LPH and β-LPH, so that these post-translational changes are regulated during development. Acetylation and glycosylation of the peptides also occur and might be important in changing the susceptibility of a peptide to degradation or in altering its biological activity (Hughes, 1979). The further analysis of these complex biochemical events in relation to the *in vivo* neuronal activity of the cells containing these peptides (where it is already known that they can elicit effects similar to the opiate drugs) provides an exciting challenge and may be of profound importance to the study of neurobiology.

Somatic cell hybrids

A completely different approach to the analysis of genetic interactions is to examine the effect of chromosomes from different cells on gene expression when they are brought together in the same nucleus in somatic cell hybrids. Fusion of different cell-types can lead to either the formation of multi-nucleate heterokaryons (chapter 3) or the cell hybrids described here, where the nuclei also fuse. Viable cell hybrids can be formed by the fusion of cells of the same or of different species. In animals, the cells may be normal, derived from a mutant animal or mutant cell culture, or malignant, and may be differentiated or undifferentiated. Usually, one of the cell lines is a permanent (transformed) cell line and fusion is usually carried out between cell lines from different species so as to facilitate recognition of the origins of the different chromosomes and the electro-phoretic forms of marker enzymes in the hybrid. The system is therefore highly artificial but may still provide useful information. In plants, protoplast fusion forms an equivalent experimental system.

Although some cell hybrids will form spontaneously in mixed culture, the recovery rate is much improved if cell fusion is promoted by a virus (such as the Sendai virus also used to form the heterokaryons described in chapter 3), or polyethylene glycol. The problem then remains of isolating the cell hybrids from heterokaryons and unfused cells. This can be done by repeated subculturing and inspection of the cultures, or less tediously by the use of a selective technique. The technique most often used depends on the fact that cell lines resistant to the drugs 8-azaguanine or 5-bromodeoxyuridine can readily be isolated because they can grow when these compounds are added to the medium. These cells lack hypoxanthine-guanine phosphoribosyl transferase (HGPRT) activity or thymidine kinase (TK) activity respectively, and therefore do not incorporate these normally toxic analogues. Littlefield (1966) realized that neither of these cell lines would be able to grow on medium containing hypoxanthine, aminopterin and thymidine (known generally as HAT medium), because the inhibitor aminopterin blocks endogenous synthesis of hypoxanthine and thymidine, while the enzyme deficiency in HGPRT$^-$ or TK$^-$ cells means that they cannot incorporate the exogenous hypoxanthine or thymidine respectively. However, when mouse fibroblasts mutant for either HGPRT$^-$ or for TK$^-$ were put together in HAT medium, hybrid cells, which produced both HGPRT and TK at about half the wild-type parental concentrations, were found to grow well (figure 6.7). The TK$^-$ and HGPRT$^-$ mutations are thus said to be "recessive" (but the term is

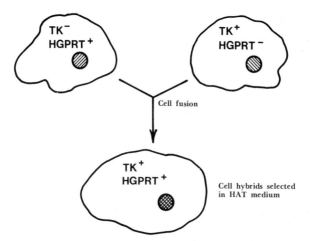

Figure 6.7   Selection of hybrid cells in HAT medium from fusion between thymidine kinase (TK) deficient and hypoxanthine guanine phosphoribosyl transferase (HGPRT) deficient fibroblasts, induced by the presence of Sendai virus.

being used in a rather different sense to the usual one where it applies to the relationship between alleles of the normal diploid cell). Sometimes, only one of the parental cell lines is mutant (TK$^-$ or HGPRT$^-$); however, the hybrid cells generally grow as well or better than the parental wild-type cells, so that this "half-selective" system is quite productive and has the advantage that it can be used with normal differentiated cells. Many HGPRT$^-$ and TK$^-$ cell lines have been isolated from cultures of mouse, rat, human, Syrian hamster and Chinese hamster cells.

Somatic cell hybrids have been used in a variety of ways. In hybrids between different rodent species, the chromosome complement of the hybrid remains fairly stable. However, in mouse × human hybrids, the various human chromosomes are lost preferentially, and more or less at random in different cell cultures. This fact is particularly useful for human linkage studies; by analysing sufficient cultures, the disappearance of the human form of a particular enzyme can be correlated with the loss of a particular chromosome in the same cultures, thus localizing a locus controlling the expression of that enzyme to that chromosome. By this means, the chromosome map of man has been dramatically extended in recent years. In addition to these important formal studies, hybrids have been used in the analysis of the chromosomal control of enzyme expression, mutant phenotypes, differentiated characteristics and malignancy. In the case of enzymes present in both parents, the hybrid continues to make both forms of the enzyme so that there is no evidence concerning the regulation of these common enzymes. However, if the enzyme is multimeric, hybrid bands of the enzyme may be seen after electrophoresis, providing evidence of molecular homology between the parental enzymes (e.g. Syrian hamster and mouse malate dehydrogenase; rat and mouse lactate dehydrogenase).

Examples of the study of differentiated functions in hybrid cells are the production of pigment in melanoma cells and the induction of tyrosine aminotransferase by dexamethasone in liver cells (see chapter 4). When pigmented Syrian hamster melanoma cells, with a high DOPA oxidase activity (responsible for the production of melanin from tyrosine), are fused with unpigmented mouse L cells (undifferentiated cells), all the hybrids are unpigmented and lack detectable DOPA oxidase (Davidson and colleagues, 1966). When hybrids with three genomes, two of them from pigmented and one from unpigmented cells, were isolated, about half were pigmented and half unpigmented. On subculturing, the pigmented clones produced some unpigmented sub-clones, but the converse were never seen. A simple model to explain these observations is to suppose that the unpigmented cell elaborates some repressor substance which inhibits DOPA oxidase formation. In the tetraploid hybrids, this inhibits DOPA oxidase production by the Syrian hamster genome; in the hexaploid hybrids, it is present in sufficient quantities to do this in only about half the cells. The opposite model—that the melanoma cells produce an activator substance which reaches a sufficient concentration for DOPA oxidase production in only about half the hexaploid hybrids—is equally possible.

In a rather similar experiment, hybrids can be made between HGPRT⁻ rat hepatoma cells (inducible for TAT) and HGPRT⁺ human fibroblasts. In these hybrids, selected in the HAT medium, the TAT inducibility is lost. However, if they are placed in 8-azaguanine medium which selects for HGPRT⁻ cells (i.e. for the loss of the human HGPRT⁺ X-chromosome), the surviving clones regain their TAT inducibility (Croce and colleagues, 1973). All these hybrids retain their steroid acceptor. The simplest explanation for this is that the loss of the differentiated function in the hybrid is due to repression by the X-chromosome of the undifferentiated cell; the hybrid nevertheless retains its determination as a liver cell and regains its differentiation (using the enzyme as a marker for a particular differentiated function) when the repressive X-chromosome is lost.

The loss of differentiated functions when a hybrid is formed is therefore the most common result, but differentiated functions are sometimes retained (e.g. excitable membranes, acetylcholinesterase synthesis and sensitivity to acetylcholine in hybrids between mouse L cells and neuroblastoma cells). It is also possible for new characteristics to appear in the hybrid. Although a number of regulatory models can be constructed to explain these results, a precise understanding of the processes involved remains elusive; conventional genetic analysis is impossible and our knowledge of the biochemical basis of chromosomal interactions is, at best, rudimentary. With these limitations, it is unlikely that significant advances in our understanding of gene interactions will come from the study of cell hybrids. Nevertheless, they do provide a unique experimental system for highlighting the extensive interactions between chromosomes that may occur during development.

Controlling elements

Controlling elements represent an intriguing concept. They were originally identified by classical genetic studies on maize (notably by McClintock and Brink) and they have also been identified in *Antirrhinum majus*, *Nicotiana* and *Drosophila*. Currently, restriction mapping and related techniques are being used to define the properties of the controlling elements. These studies reveal them to be analogous to the transposable elements in prokaryotes (insertion sequences and transposons). The functional significance of controlling elements remains controversial.

The literature on controlling elements in maize is widely regarded as difficult and complicated, but it has been reviewed by Fincham and Sastry (1974). The most convenient markers are mutations affecting the pigmentation of the triploid endosperm (which develops from the fusion of two maternal gametic nuclei and one gamete from the pollen tube); the phenotype of the endosperm generally reflects the potential phenotype of the underlying embryo.

A controlling element, also known as a transposable element, is recognized by the two characteristics implied in these names: an effect on gene expression (or activity) and an ability to move to different sites in the genome. In general, a controlling element is postulated to inhibit the effect of an adjacent (*cis*) allele. If an inactive allele becomes active, then a controlling element may have been removed from its adjacent site since back-mutation to an active allele, the alternative explanation, is exceedingly rare. If the activity of a previously active allele at another locus is inhibited simultaneously, or if the phenomenon is regularly observed with the unstable allele repeatedly giving rise to a stable active allele, then conventional point mutation becomes an inadequate explanation. The apparent genetic instability (or mutability) of an allele due to its association with a controlling element may be seen in either somatic tissue (leading to some form of spotting or variegation) or in germinal tissue (where the change is heritable and appears as a new mutant form of the allele). In many cases the controlling element is still sufficiently stable to be mapped at various positions around the genome as it moves and is integrated near to a particular allele, affecting its activity, before being excised again and re-inserted elsewhere. However, it could apparently be lost if it were inserted next to an allele without a recognizable phenotypic effect on the endosperm.

The simplest type of controlling element, as outlined above, only requires consideration of the one element which affects the alleles adjacent to its chromosomal location, and is therefore known as a one-element (or autonomous) system. Quite often, two elements have to be considered: the one next to the affected allele is stable unless a second element, located elsewhere in the genome, is also present. Thus the second element, termed the *regulator element*, seems to supply some function necessary for the excision of the first element (termed by McClintock the *operator element*;

Fincham and Sastry suggest the term *receptor element* as less liable to suggest an over-precise analogy with the bacterial operator).

At least three classes of controlling elements can be identified because receptor elements of one class respond only to regulator elements in the same class, and not to other regulator elements. Within each class a variety of different forms of both regulator and receptor elements can be found with different properties. These classes are known as *Dotted (Dt), Activator (Ac)* and *Suppressor-mutator (Spm).* As indicated by its name, *Spm* has two functions which can be separated in different forms of the element. The *Spm* receptor element alone causes some degree of repression of the adjacent allele (*cis*-acting); if the regulator element is introduced by an appropriate cross, this suppresses gene activity completely in most cells (the *suppressor* function) and causes excision of the receptor, with the restoration of the full activity of the adjacent allele, in others (the apparent *mutator* function). The three classes of elements also show different dosage effects on the frequency of excision (apparent mutation); *Spm* has an all-or-none effect depending on its presence or absence, the effect of *Ac* is maximal with one copy present and reduced progressively by the introduction of a second and a third into the triploid endosperm, whilst the effect of *Dt* increases in proportion to the number of copies present.

It can be argued that the two element systems are derived from autonomous elements. For example, the a_1^{m-2} allele has closely linked *Spm* activity and produces a variegated pattern of dark spots on a white background presumably due to the effect of the integrated controlling element (McClintock, 1962). Occasionally, this can give rise to a pale mottled mutant allele of a_1 lacking *Spm* activity; if *Spm* is introduced elsewhere in the genome by a suitable cross, then this new allele produces the normal a_1^{m-2} variegated phenotype. This could be explained if the pale mottled allele arose from the *Spm* a_1^{m-2} allele by the loss of regulator function from *Spm*, leaving the receptor function intact to produce the phenotype of the new allele. When the *Spm* regulator function is present elsewhere in the genome, the receptor would respond to it to give the a_1^{m-2} phenotype again. Thus, an autonomous element (*Spm* adjacent to the a_1 locus) has given rise to a two element system (with only the *Spm* receptor adjacent to the a_1 locus).

The controlling elements described so far are important as one explanation for the development of a mottled or variegated phenotype. Controlling elements can also cause a regular change in pigment expression, leading to consistent patterns characteristic of the form of the controlling element. For example, Peterson (1966) found two different derivatives of the a_1^m allele which produces pigmented spots due to the presence of *Spm*. The first derivative, $a_1^{m(crown)}$ produces pigmented spots only on the crown of the kernel, opposite its insertion of the cob; the second derivative, $a_1^{m(flow)}$ produces pigmented spots round the waist of the kernel. Each of these new alleles was combined with the a_1^{m-1} allele in the heterozygote. Normally, a_1^{m-1} is non-autonomous and requires *Spm* as a regulator element; in the absence of *Spm*, it produces uniform intermediate pigmentation; in the presence of *Spm*, dark spots on a white background are seen. In the $a_1^{m(crown)}/a_1^{m-1}$ heterozygote, the crown had pigmented spots reflecting an action of the a_1 *Spm* regulator element on both alleles, but the rest of the kernel was now uniformly pigmented showing an absence of *Spm* regulator activity on a_1^m in this region. Likewise, in the $a_1^{m(flow)}/a_1^{m-1}$ heterozygote, the waist was spotted as before, but the crown was uniformly pigmented. The *Spm* elements of the new alleles, $a_1^{m(flow)}$ and $a_1^{m(crown)}$ derived from the original a_1^m allele therefore show controlled changes in activity in relation to the position of the cells in the kernel. Since position reflects a difference in the time of cellular differentiation between

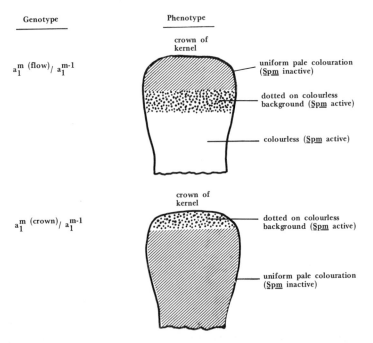

Figure 6.8   Pigmentation patterns in corn kernels.

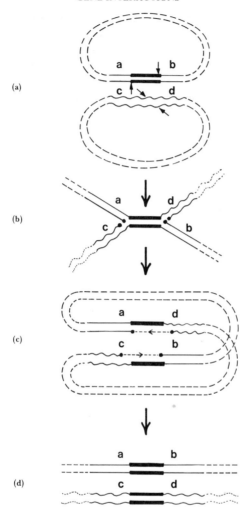

Figure 6.9 A model for the replication of transposable elements. (In this example, circular genomes are shown but this is not a necessary feature of the model).

(a) Single-stranded cuts at the ends of the transposable element and staggered cuts of opposite polarity in the target site are made.

(b) Each cut end of the transposable element is attached to a protruding end of the staggered cut, creating two replication forks.

(c) Semi-conservative replication generates two new transposable element strands, and short repeat strands (in the same (direct) orientation) of the region between the two staggered cuts.

(d) Crossing-over at a specific site within the direct repeats or transposable element resolves the structure into the original duplices, *ab* and *cd*. (After Shapiro).

the crown and the waist, the different forms (or states) of the *Spm* element in the *crown* and *flow* alleles must be activated at difference times during development, resulting in the observed pigmentation patterns (figure 6.8). How this time sequence of activation is regularly achieved is not known.

In parallel with the maize controlling elements, the existence of transposable elements in other eukaryotes can be inferred by their effects on gene activity. For example, highly mutable chromosomes (MR chromosomes) have been identified in wild-caught *Drosophila* (Berg, 1974; Hiraizumi, 1971) by their production at high frequency of unstable mutants at a number of loci when bred in the laboratory. Movable genes (transposons) have also been identified in *Drosophila*, and although they are restricted to chromosomal segments containing the *white* eye gene, integration can occur at a large number of sites scattered throughout the genome (Ising and Ramel, 1976).

At the molecular level, transposable elements in eukaryotes seem to be homologous to the so-called insertion sequences of prokaryotes which are found in bacteriophages, and in such plasmids as the F fertility factor and the R factors, transmissable plasmids mediating drug resistance. When these are transposed, insertion is associated with the production of a short repeated sequence in the DNA (see Calos and Miller, 1980). Insertion is not, however, accompanied by excision elsewhere (figure 6.9).

In eukaryotes, transposable elements can be identified by Southern blotting of restriction enzyme digests of chromatin as repeated DNA sequences recovered at different positions in closely-related strains. Five such repeated families of sequences designated 412, *copia*, 225, 234 and 297 (Green, 1980) have been identified in *Drosophila*. Such sequences code for

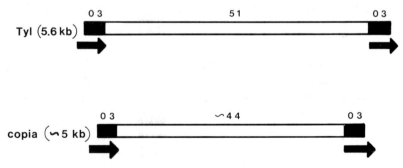

Figure 6.10 The structure of two eukaryotic transposable elements. Ty1 is present in about 35 copies in the yeast genome, copia about 35 times in *Drosophila*. Both types of sequence are flanked by a short (0.3 kb) repeated sequence in direct orientation (arrowed). kb—kilobases.

an abundant mRNA but may be found at different chromosomal sites in different independently-isolated wild type strains (Strobel and colleagues, 1979), and even in different individuals of a single strain (Ilyin and colleagues, 1978). A *copia* sequence is responsible for the transposition of the *white* gene described above (Gehring and Paro, 1980). An insertion sequence (Ty1) present in about 35 copies per genome has been identified in yeast (*Saccharomyces cerevisiae*). Significantly, both the Ty1 sequence in yeast and the *copia* sequence in *Drosophila* (see Calos and Miller, 1980) are associated with a short sequence repeated at either end of the insert (figure 6.10); in parallel with prokaryote insertion sequences, transposition and sequence insertion in eukaryotes may lead to the duplication of the target sequence.

Mating type of yeast

In yeast, the mating type of a or α depends on the expression of either a or α regulatory information at the mating type (MAT) locus. High frequency changes in mating type are seen in homothallic strains carrying the HO gene; interconversions of mating type may occur with every cell division. Other heterothallic strains carrying the ho gene are considerably more stable with interconversion frequencies of 10^{-6}. These changes in mating type are thought to involve the transposition of genetic information (a cassette) from silent libraries of a and α information at two genetic sites, HMa and $HM\alpha$, located on the left and right hand arm of chromosome three respectively (Harashima and Oshima, 1976) of all cells to the MAT locus where it is expressed (figure 6.11). The evidence for this model comes largely from two types of observations. Firstly, mutations at the MAT locus can be corrected or "healed" by mating type interconversion (Hicks

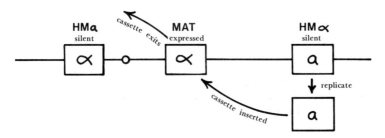

Figure 6.11 The cassette model for mating type switching in yeast. In this example, a switch from $\alpha \rightarrow a$ is being made. (Redrawn from Hicks and colleagues).

and Herskowitz, 1977) and secondly, mutations at HMa and $HM\alpha$ can move to MAT and be expressed as phenotypes of known mating type mutations (Blair and colleagues, 1979). Switching involves the transfer of a new cassette of regulatory information to MAT (to give conversion of a to α or α to a), with the loss of the previous cassette at MAT. Variants of the silent libraries have been found: hma which prevents the conversion of a to α, and $hm\alpha$, which prevents the conversion of α to a. However, hma and $hm\alpha$ are functionally equivalent to $HM\alpha$ and HMa respectively, since HMa, $HM\alpha$ and hma, $hm\alpha$ strains show switching in both directions.

Developmental role of controlling elements

In looking for a developmental role for controlling elements and transposable sequences, there are several features that should be considered. The first point is that although the investigations described have been carried out in only a few species, this has been primarily for technical reasons and it seems likely that such sequences are present in other species. For example, an inserted sequence could account for the *pink-eye-unstable* mutant in the mouse which regularly produces a mosaic of dilute and intense pigmentation in the coat (Melvold, 1971). In maize, new controlling elements appear when a "sticky" chromosome is subjected to a repeated cycle of fusion and breakage during cell division, suggesting that they were previously integrated into the chromosome in a more stable fashion. If it is accepted then that such sequences are widespread, what functions do they carry out? There are several possibilities.

1. As in prokaryotes, there may be a relationship between controlling elements and viruses. Many viruses are in inverted or direct repeats; RNA tumour viruses are integrated in the genome as a DNA sequence flanked by direct repeats of 600 base pairs, similar therefore to the structure of transposable elements in figure 6.10. Mouse tumour viruses such as mammary tumour virus, sarcoma virus and leukaemia virus also give rise to the short direct repeat (4–5 base pairs) on insertion which is characteristic of transposable elements. These sequences could therefore be a form of parasitic DNA (see Doolittle and Sapienza, 1980).

2. It could be argued that transposable elements play an evolutionary rather than a developmental role. Their relative lack of specificity in integration sites means that they may mobilize almost any sequence, and bring it adjacent to any other sequence. Where a protein has two recognition sites or domains (e.g. an enzyme with its active site which is

also subject to allosteric inhibition via a second site), the coding sequences of the two domains could have evolved separately in the genome and subsequently been brought together in this way (see p. 73). The introns in coding sequences could be the remnants of the transposable elements; they are usually flanked by imperfect direct repeats which could have arisen by mutation from the direct repeats of transposable elements.

3. Transposable elements may be involved in the rearrangements of the genome that occur during development. In the immune system, the joining of different V, J and C genes (see page 17) is partly responsible for generating the diversity of immunoglobulins seen in individuals; this is achieved by the re-arrangement of the DNA of V and J coding regions to give a single functional unit for the production of immunoglobulin mRNA. Transposable elements with some degree of specificity would have the necessary properties to carry out these transpositions. Transposition of genetic information is also seen in mating type switching in yeast. Normal development could therefore be accompanied by changes in gene expression arising from the regular movement of controlling elements within the genome, or to regular changes in the state of the controlling elements as seen in the pigmentation patterns of the *crown* and *flow* alleles of maize. The evidence discussed in chapter 2 concerning the constancy of the genome does not resolve the issue, because changes in the site of controlling elements would not be detected by the techniques of nucleic acid hybridization of whole genomes. Furthermore, the totipotency of the genome could be explained if the changes in site or state of the controlling elements were reversible. More detailed molecular studies, including the use of restriction mapping techniques, will be required to test for this possibility.

These three possibilities are not of course mutually exclusive, and the different characteristics of a given controlling element could be responsible for its continued existence at different times in the evolution of the organism. Indeed, the different categories must be related, because a controlling element could not produce evolutionary flexibility, however desirable, in the long term unless it is maintained in the population either as a parasite or by virtue of some developmental role; conversely, if a controlling element is present in the population, then it can clearly evolve a new role, either as a self-perpetuating sequence without a developmental role or in the direction of a specialized function. Introns with a regulatory role derived from transposable elements could be one example of this.

Whatever the balance between these different roles proves to be, controlling elements are changing the view of the genome as a stable entity towards a concept where it may be much more fluid, both in development and in evolution.

Conclusion

In this chapter, we have seen how genes at many loci can affect the expression of a developmental defect, either by modifying the expression of a single mutant gene or by their cumulative effects on the canalization of the character. In both cases, the end result of abnormal development is the same. Chromosomal interaction may also occur, as seen in cell hybrids, with many differentiated functions being reversibly lost after the fusion of differentiated and undifferentiated cells. The situation is however extremely complex, since differentiated functions are not always lost, and the effects can rarely be attributed to single chromosomes, let alone individual gene loci.

The analysis of complex loci provides a different approach to the problem of gene interaction in development. In these cases, the chromosomal position of the gene is important, suggesting that development may proceed by the sequential activation of genes on the chromosome. Additionally, as in the case of the *r* allelic series in *Drosophila*, the interaction of alleles may arise from the production of a single multi-enzyme complex. Complex loci reflect a particular type of chromosomal organization where genes of related function are collected into genetic units, either as a result of gene duplication or sequence rearrangement during evolution. Controlling elements provide a mechanism for the rearrangement of gene sequences both in evolution and during development. The more general occurrence of gene transpositions in the processes of cell determination and differentiation must remain an exciting possibility.

CHAPTER SEVEN

THE ORGANIZATION OF DEVELOPMENT

At the cellular level, development proceeds by the selection and activation of a particular developmental programme involving the sequential action of a number of gene loci. As the programme unfolds, groups of cells become destined to give rise to a particular tissue or structure. At what stage can they be identified and what is the usual origin within the normal embryo of these cells? This is essentially the question of cell lineage, of defining the sequence in which cells within the embryo adopt their presumptive fates. An understanding of cell lineage is a necessary preliminary to an investigation of the related problem of mechanism. What are the influences which bring about the changes in the cells as they achieve their presumptive fates? Are they intrinsic to the cells at the earliest time they can be identified (autonomous development), or do they depend on interaction with other cells or cell products at some later period?

Chapter 3 included a discussion of whether the egg contained localized autonomous instructions (the mosaic theory), or whether influences from outside the cell were acting to determine its development, as in the reference point theory. In the mosaic egg, or once a cell has become determined, development is necessarily defined by cell lineage. On the other hand, in the reference point or regulative egg, cell lineage defines only the presumptive fate of the cell; its actual determination and differentiation depend on interaction with other regions—the reference point in the egg, or the local process of embryonic induction between adjacent tissues. However, neither determination nor differentiation can be thought of as single events occurring at a single point in the development of the cell or tissue. Rather, they involve a sequence of events that progressively restrict the developmental potential of the cell, and culminate in overt differentiation. Even then, regeneration in certain species indicates that determination and differentiation may be reversible to some degree. For example, following the surgical removal of the newt eye lens described in chapter 2, the iris dedifferentiates and forms a new lens. In *Ilyanassa* development, the removal of the polar lobe from the cleavage-stage embryo (p. 34) results not only in the failure to form cells normally derived from that lineage (which can be interpreted as a requirement for that cytoplasmic determinant for autonomous development) but also in the failure of development of mesodermal derivatives of a different cell lineage which require an interaction by embryonic induction with the cell products of the polar lobe for their non-autonomous development. Thus, the problems of cell lineage and autonomy need to be considered throughout all phases of development.

The experimental methods used to investigate these problems are relatively straightforward in principle. To investigate cell lineage, various cells within the embryo can be marked—for example, with vital dyes, carbon particles, radioisotope labelling of DNA, or some genetic marker which is thought to be autonomous in expression (e.g. a chromosomal translocation)—and the destination of these cells can be followed on the assumption that the marker does not interfere with development. Such experiments are designed therefore to ascertain the fate of the marked cells. By contrast, to test whether development is autonomous or not, cells are deliberately exposed to unusual developmental environments. For example, they may be isolated from other cells, grafted to another site in juxtaposition with other tissues which might influence them, or their development in proximity to cells of a different genotype observed following either grafts between genetically-distinct individuals or the formation of genetic mosaics or chimaeras. This type of experiment defines operationally the stage at which the cells become determined and capable of autonomous development to a particular cell-type or a restricted range of cell-types.

This chapter describes some genetic techniques used for determining cell lineage and autonomy. The use of genetic variants in the analysis of the mechanisms which operate in interactions between cells and tissues is considered in the next chapter.

Genetic mosaics

The genetic analysis of cell lineage requires the co-existence within a single individual of two or more distinct cell populations. Mixed cell populations may occur naturally, or be produced artificially. The term *mosaic* is used to denote an individual in which two cell populations have arisen during development by such mechanisms as chromosomal non-disjunction or somatic recombination (or even by X-inactivation in mammals) from a single zygote. The term *chimaera* is used where the individual results from a mixture of cells derived from two (or more) separate zygotes.

Chimaeras in the mouse
Mintz (1962) and Tarkowski (1961) independently developed a technique for the production of mice with cells from two different embryos. These chimaeras are also referred to as allophenic or tetraparental mice (reflecting the origin of the two zygotes from four parents). Early cleavage-stage embryos are removed from their mothers and the zona pellucida removed by incubation in a medium containing the enzyme pronase. The cells of the two embryos may then be aggregated to give a single over-size embryo. The composite embryo is cultured for a period before being injected into the uterus of a pseudopregnant foster-mother, where a proportion of them develop into normal-size embryos and eventually into fertile adults (figure 7.1). Gardner (1968) developed an alternative technique which achieves the same end, whereby a single cell from a cleavage-

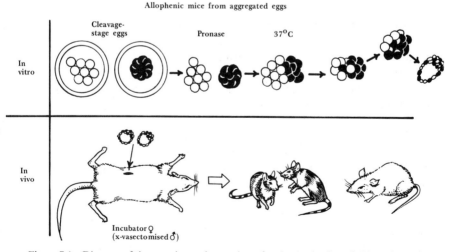

Figure 7.1 Diagram of the experimental procedures for the production of chimaeric mice. Blastomeres of the two contributing strains of differing coat colour are shown in black and white. (Redrawn from Mintz).

stage embryo is micro-injected into a blastocyst; if it forms part of the inner cell mass, it can then contribute a large proportion of the cells of the foetus. The formation of chimaeras can be confirmed by the use of appropriate genetic markers to distinguish the two contributing cell types—at its simplest, by using embryos which would form mice of different coat colours (and may also be distinct from the coat colour of the foster-mother).

The use of these techniques in the study of cell lineage requires locally acting markers to distinguish the contributions of the two cell lines. Suitable markers are scarce: in addition to coat colour genes (which are not always completely cell autonomous), enzyme markers which can be visualized in intact tissues either by histochemical techniques (e.g. β-glucuronidase in strains with high and low enzyme activities) or by examination of tissue fragments for electrophoretic variants, and aberrant karyotypes which can be examined in particular cells (e.g. the T6 translocation) have been most widely used.

Mintz and Baker (1967) used chimaeras to investigate the origin of the multinucleate myotubes of striated muscle. These muscle cells could arise from myoblasts, which have only a single nucleus, either by repeated division of the nuclei without cell division, or by the fusion of different cells. Embryos from two strains carrying distinct electrophoretic variants of the enzyme isocitrate dehydrogenase were combined. The parental strains produce only a single band on electrophoresis of tissues such as liver or muscle, whereas F_1 hybrids produce not only these two bands but an additional intermediate hybrid band, indicating that the active enzyme is composed of two polypeptide subunits (i.e. it is dimeric). In some tissues of the chimaeric mice such as liver, both parental forms of the enzyme were produced, but no hybrid bands, indicating within the limits of sensitivity of the technique that (a) the animals were chimaeras, and (b) the expression of the form of the enzyme is cell-autonomous. In contrast, skeletal muscle contains both the parental and hybrid bands. The production of hybrid enzyme will occur only when the alleles coding for the two subunits are present in the same cell, as in conventional heterozygotes or in cells containing genetically-distinct nuclei. Such heterokaryotic cells can of course arise only by cell fusion. Therefore, unless this enzyme marker is behaving in an aberrant non-autonomous manner in skeletal muscle tissue, the multinucleate myotubes must be formed by cell fusion.

A second example of the use of chimaeras is in the study of pigment cell lineage in the development of the pigmentary system of the coat. Early analyses tended to assume that the number of different coloured patches corresponded to the number of cell clones making up the tissue. In general, this is unlikely to be true because adjacent clones will sometimes be of identical genotype and hence produce the same colour. Conversely, depending on the rate and direction of cell proliferation, as well as on cell mingling, a single clone may become divided into two or more patches. Models investigating the consequences of this have been developed for the one- and two-dimensional cases by West (1975) and by Ransom and colleagues (1975). To simplify the present discussion, let us consider the one-dimensional case, and assume that both components contribute equally to the chimaera. The average size of a patch of tissue of one type (ignoring the possibility of cell movement and mingling of some of the clonal descendants of the founder cells) will therefore be two clones. (If the two components contribute unequally to the particular tissue, as can readily occur by chance during development, then the average patch size increases markedly with increasing inequality and the total number of patches will decrease.)

An approximation to the one-dimensional situation can be found in the pigmentation pattern of chimaeras arising from the fusion of an embryo homozygous for a mutant genetic marker acting in the melanocytes (e.g. *albino* or *brown*) with an embryo carrying the corresponding non-mutant alleles. These melanocyte chimaeras (figure 7.2) give rise to characteristic

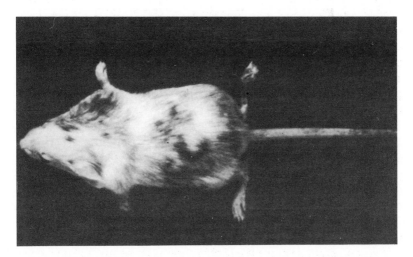

Fig. 7.2 Melanocyte chimaera. Note the pigmented (normal) and unpigmented (mutant) patches in the coat. (From McLaren (1976) *Mammalian Chimaeras*, Cambridge University Press, reproduced with permission).

patterns with broad transverse stripes from head to tail; this orientation arises from the transverse migration of the melanoblasts from the neural crest between 8 and 12 days of gestation. Some bands show cell mingling, especially towards the edges, which could result from some limited longitudinal movement of the melanoblasts during their migration. Individual hairs can also show variegation, indicating that the progeny of more than one melanoblast can colonize an individual hair follicle. Ignoring these difficulties, the basic striped pattern can be taken as reflecting the one-dimensional arrangement of primordial melanoblasts arising from the neural crest. There is a sharp discontinuity in the pattern along the mid-dorsal line, so that the pattern on the two sides is independently determined. Mintz (1970a) suggested that this could be due to the determination of pattern before the neural folds fuse, so that primordial melanoblasts do not pass from one side to the other. She found that the observed range of patterns could be explained if there were 17 primordial melanoblast regions on each side (three on the head, six on the body and eight on the tail). Thus, on the one-dimensional model outlined in the previous paragraph, there should be an average of 17/2 (or $8\frac{1}{2}$) patches (or stripes in this case) on each side. This agrees well with the average value of 8.14 determined by McLaren and Bowman (1969). It does not necessarily follow that each region is in fact colonized by the descendants of a single primordial melanoblast, and, indeed, stripes of mixed colour such as would be observed if one region were colonized by a mixed clone derived from two or more melanoblasts are seen occasionally.

The clonal origin of the photoreceptor layer of the retina has been successfully analysed with the use of the *retinal degeneration* (*rd*) mutant (Mintz and Sanyal, 1970). In strains homozygous for *rd* (e.g. the C3H strain) the visual cells form normally but degenerate post-natally. Degeneration also occurs in the *rd/rd* cells in the retinas of *rd/rd* ↔ +/+ chimaeras, leaving "ghost" areas devoid of rod elements. By the analysis of a large number of such chimaeric retinas, 10 sectors can be identified (figure 7.3). Thus the retina of each eye is apparently derived from 10 progenitor cells, located in a small circle, that proliferate radially to produce 10 clones of cells. This interpretation depends however on the absence of cell migration between adjoining sectors. The demonstration of the colonization of *rd/rd* regions by +/+ cells in chimaeric retinas by West (1976a) casts some doubt on the precision of this estimate.

In many cases, it is not possible to directly determine the spatial distribution of patches of the different cell types. Nevertheless, by the use of appropriate genetic markers (e.g. electrophoretic or chromosomal

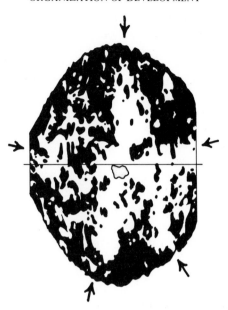

Figure 7.3 Clonal origins of the photoreceptor cell layer of the mouse retina. The drawing is a two-dimensional reconstruction from serial sections of the retina of a $rd/rd \leftrightarrow +/+$ chimaera with approximately equal numbers of $+/+$ and rd/rd cells. "Ghost" areas from the death of rd/rd cells are shown in black. Note the stellate pattern (arrows point to five $+/+$ components). (Redrawn from Mintz and colleagues).

variants) and by making certain simplifying assumptions, it is still possible to obtain an estimate of the number of cells contributing to a tissue. If the two cell types are randomly mixed at the time of tissue formation, if there is no differential cell adhesion, and if there is no selective difference between the two cell components (i.e. in the relative rates of cell proliferation and cell death), a tissue completely homogeneous for one component will become progressively rarer as the number of cells contributing to that tissue increases (according to the binomial distribution). Specifically, if only one cell is involved, the tissue must be of one type or the other; if it originates from two cells, it will be homogeneous in half of the cases; if it originates from n cells, it will be homogeneous in proportion to $1/(2^{n-1})$ of the cases. This approach has been applied to the study of development in overt chimaeras between the C3H and C57BL strains by Mintz (1970a). 71 % of such mice showed chimaerism in the hair follicles but only 44 % in the liver (table 7.1). By making some allowance for the lack of complete sensitivity of the markers, Mintz inferred that the foetus is derived from a

Table 7.1 Mosaicism amongst 129 C3H ↔ C57BL/6 chimaeras (data from Mintz, 1970a)

	% mosaics	Estimated number of initiation cells
Whole animal	70	3
Liver	44	2
(malate dehydrogenase variants)		
Hair follicles	71	3
(*agouti* v. *non-agouti*)		

region with three cells, as are the hair follicles, and the liver is derived from a region of two cells.

An alternative approach is to look at the correlation between the proportion of the two cell types in different tissues as it varies by chance over a number of individuals. If two tissues always have a very similar proportional composition, then they must be derived from a common precursor population, whereas if the two tissues are independent in their composition, then their developmental origins are unlikely to be closely related. Making similar assumptions to those outlined above, a cell-lineage may be derived from the degree of correlation between the various tissues (figure 7.4).

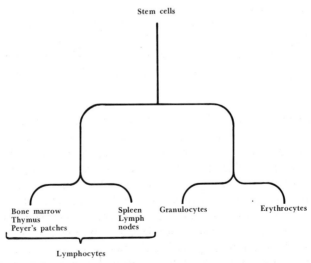

Figure 7.4 Cell lineage of lymphocytes, granulocytes and erythrocytes as derived from correlation analysis of electrophoretic variants in chimaeras and X-inactivation mosaics.

As mentioned previously, such estimates are acceptable only if certain conditions can be met. In particular, coherent clonal development after the determination of cell lineage must occur if patch size is to be used as a criterion for estimates of primordial cell number. This has been shown not to be the case in the development of the pigmented layer of the retina (West, 1976b) where considerable cell mingling occurs until $12\frac{1}{2}$ days of gestation. Considerable cell mixing is also seen in the liver (West, 1976c). Cell selection may also occur and this may not be the same in different tissues. For example, in the C3H $\leftrightarrow$ C57BL chimaeras studied by Mintz (1970b), the kidney, liver, mammary glands and erythropoietic tissues each have different average proportions of the two cell types. In an attempt to overcome these difficulties, Mintz (1970b) has developed the concept of the Statistical Allophenic Mouse, or Sam, each of whose tissues has the most likely proportions of the two cell types. The binomial distribution can then be corrected for the combined effects of differential viability and proliferation seen in particular strain combinations.

In summary, therefore, the analysis of cell lineage in many tissues of mammalian chimaeras has encountered serious difficulties, particularly as a result of extensive cell mingling and the differing competitive success of cells of different strains in the different tissues of chimaeras. A further limitation is that the formation of chimaeras is restricted to the mid-cleavage stage of development, whereas the distribution of cell types will, in many cases, be determined in more adult tissues. As we have seen, any inequality in the numbers of the two cell types must necessarily be magnified by this stage. For an example of a rather different approach to the problem of early cell lineage in the mouse, involving sequential studies of X-inactivation phenotypic mosaics rather than chimaeras, see p. 91.

Mosaic mapping in *Drosophila*: chromosome loss in early cleavage

As outlined in chapter 3, the first nuclear divisions in the *Drosophila* embryo occur without the formation of cell membranes. After eight divisions the nuclei migrate to the periphery of the egg cytoplasm and a small group of pole cells forms at the posterior pole. The other nuclei form a monolayer, and after an additional 3 or 4 synchronous divisions, cell membranes are formed to produce the surface blastoderm layer (see figure 7.6). The larval and imaginal tissues develop from this blastoderm layer. The imaginal tissues develop during the pupal period from undifferentiated cells present throughout larval life as discrete groups of small epithelial cells or imaginal discs arranged along the body of the larva.

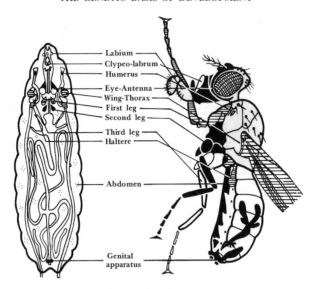

Labium
Clypeo-labrum
Humerus
Eye-Antenna
Wing-Thorax
First leg
Second leg
Third leg
Haltere
Abdomen
Genital
apparatus

Figure 7.5 Imaginal discs and corresponding adult structures of *Drosophila*. (Redrawn from Nöthiger).

These cells divide throughout larval life but fail to differentiate until pupation. When the larva pupates, the imaginal disc cells undergo extensive proliferation and, with the histolysis of most of the larval tissues, differentiate into the structures of the adult fly (figure 7.5). The adult is formed therefore rather like a jig-saw from the differentiated products of pairs of imaginal discs (e.g. the head is formed from a left- and right-hand disc by fusion in the mid-line). The largely separate origin of these components makes them ideal for the study of cell lineage by mosaic mapping techniques.

In 1929, Sturtevant discovered that in *Drosophila simulans*, eggs produced by females homozygous for ca^{nd} (claret-nondisjunction, an autosomal mutant) often lost a chromosome by nondisjunction during the first or second cleavage divisions. A similar mutant was subsequently identified in *D. melanogaster* (Lewis and Gencarella, 1952). If the chromosome lost was an X or fourth chromosome, then the egg was still viable and the offspring was an XX/XO mosaic or a haplo-4 mosaic. The *paternal loss* mutant of *D. melanogaster* also leads to chromosome loss at the first or second cleavage divisions, although in this case, only paternal chromosomes are lost, whereas the *mitotic loss inducer* mutant results in loss of both paternal and maternal chromosomes but at the third or fourth

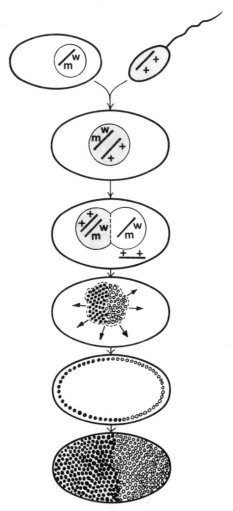

Figure 7.6 The formation of an XX/XO mosaic by chromosome loss in first cleavage. If the X-chromosome is marked with suitable alleles (w—*white eye*, m—*miniature wing* in this example) the two clones of cells will give rise to different phenotypes in the adult.

cleavage divisions. The best method for generating genetic mosaics involves the use of an unstable ring-X chromosome (In{1}w^{vc} or R{1}5A). Loss of the ring-X occurs reproducibly at early cleavage, and the mosaicism can be followed by the incorporation of autonomous cuticular marker genes, such as *yellow* (pale bristles and body colour), *forked* (bristles twisted) and *multiple wing hairs* (two to five hairs instead of one), on to the normal rod-X. XO regions can then be identified by the expression of the mutant phenotypes (figure 7.6). The XO regions of the fly are also male in phenotype, indicating that the expression of sex is cell-autonomous in *Drosophila* (such XO/XX flies are called gynandromorphs). The demarcation between the XO and XX halves of the fly may occur in any direction (i.e. horizontally, longitudinally, transversely or diagonally); this reflects a more or less random orientation of the mitotic spindle at the first cleavage divisions (figure 7.7). The demarcation line is also sharp, indicating that there is little cell mingling during the development of the imaginal discs.

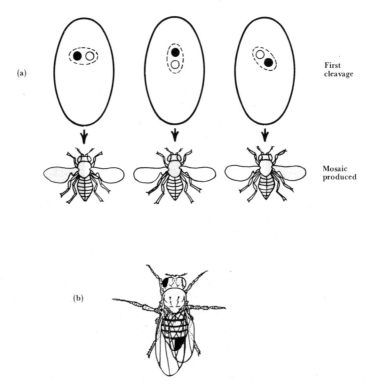

(a)

First
cleavage

Mosaic
produced

(b)

Figure 7.7   (a) The effect of the angle of the first nuclear cleavage when X-chromosome non-disjunction occurs to generate XX/XO gynandromorphs in *Drosophila*.
  (b) A bilateral gynandromorph, derived from a $+ +/wm$ heterozygote as in figure 7.6. The left half of the fly is wild-type female, the right half *white-eyed miniature* male. (Modified from Suzuki).

The technique of fate mapping developed by Garcia-Bellido and Merriam (1969) takes advantage of this lack of cell mingling to identify the relative positions of prospective imaginal disc cells within the blastoderm. The closer together in the blastoderm the cells are which will form two distinct adult structures (e.g. wing and head), the smaller the chance that the demarcation line between the two marked populations of cells in any given fly will fall between them. In other words, the percentage of mosaics where the two structures are differently marked can be taken as a measure of the distance between the two cell populations in the blastoderm. The map units used for this measure of distance have been designated *sturts*, in recognition of the contribution of A. H. Sturtevant to the development of this type of analysis. If the distance of each structure is measured from a third (e.g. proboscis), then a map of the regions of the blastoderm containing the progenitor cells for a particular adult structure can be constructed (figure 7.8).

It is also possible to map the blastoderm origin of cells in which a cell-autonomous physiological defect is expressed (Hotta and Benzer, 1972). For example, flies homozygous for the *drop-dead* (*drd*) mutant literally "drop dead" a few days after eclosion from the puparium. By constructing a rod-X carrying the *drd* mutant together with the cuticular markers *y* and *f*, and the eye colour mutant vermilion (*v*), the expression of the *drd* phenotype can be followed in XX/XO flies in relation to the expression of the surface marker genes. The association of *drd* expression with the production of a particular marked cuticular structure will identify the region of the blastoderm which gives rise to tissues expressing the *drd* defect. As shown in table 7.2, *drd* is most closely associated with the expression of *y v f* in the head region. In fact, Hotta and Benzer have located the focus of action of *drd* to the site of origin of the adult brain tissue in the blastoderm (figure 7.8).

The analysis of XO/XX mosaics has also been used to estimate the number of progenitor cells initially set aside to give rise to a particular adult structure. Postlethwait and Schneiderman (1971) found that the extent of antennal mosaicism could be divided into nine or ten classes of increasing amounts of mutant marked tissue (figure 7.9). If it is accepted that these classes identify antennae in which the XO region of the blastoderm includes progressively more of the antennal progenitor cells, then it can be inferred that the complete antenna contains a maximum of ten clones of cells. Similar estimates have been made for other adult structures—for example, each leg contains 20 clones and the wings are each derived from 12–20 precursor cells.

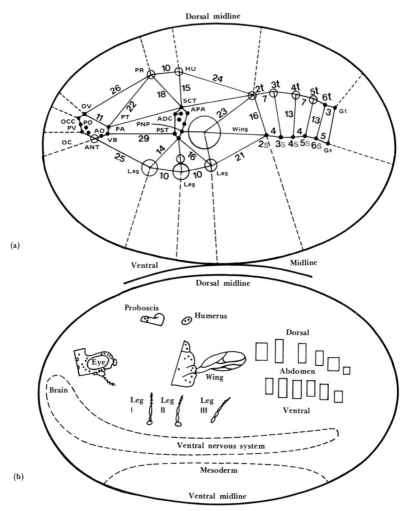

Figure 7.8 (a) Fate map of *Drosophila* blastoderm constructed by mosaic mapping. Some of the relevant distances are given in sturts. The size of circle used to show a site is proportional to the frequency with which it is split by the mosaic boundary.

ADC—anterior dorsocentral bristle	OCC—occiput	PST—presutural bristle
AO—anterior orbital bristle	OV—outer vertical bristle	PT—postorbital bristles
ANT—antenna	PA—palp	PV—post-vertical bristle
APA—anterior postalar bristle	PNP—posterior notopleural bristle	VB—vibrissae
HU—humeral bristles		t—abdominal tergites
OC—ocellar bristle	PO—posterior orbital bristle	s—abdominal sternites
	PR—proboscis	G—genital

(b) External components of the adult shown on the blastoderm surface area from which they arise. Dotted lines indicate areas identified by traditional embryological studies. (Redrawn from Hotta and Benzer).

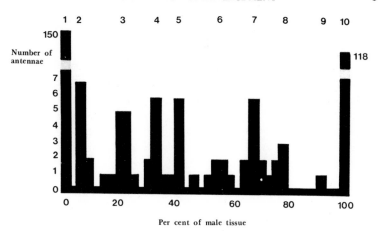

Fig. 7.9 The percentage of male tissue in antennae of gynandromorphs, showing nine or ten peaks. (Redrawn from Postlethwait and Schneiderman).

Table 7.2 Mosaic mapping of drop-dead (*drd*) behavioural focus (from Hotta and Benzer, 1972).

Head

Survival behaviour	Entire head cuticle	
	Normal $(y^+ v^+ f^+)$	Mutant $(y\, v\, f)$
Normal (drd$^+$)	91	8
Mutant (drd)	6	72
Distance of focus from head = 14/177 = 8 sturts		

Thorax

Survival behaviour	Entire thorax cuticle	
	Normal $(y^+ v^+ f^+)$	Mutant $(y\, v\, f)$
Normal (drd$^+$)	51	12
Mutant (drd)	6	47
Distance of focus from thorax = 18/116 = 16 sturts		

Abdomen

Survival behaviour	Entire abdomen cuticle	
	Normal $(y^+ v^+ f^+)$	Mutant $(y\, v\, f)$
Normal (drd$^+$)	54	28
Mutant (drd)	23	31
Distance of focus from abdomen = 51/136 = 38 sturts		

Mosaic mapping: mitotic crossing-over in *Drosophila*

The resolution of mosaic mapping in XO/XX gynanders is limited by the restriction of chromosome loss to early cleavage divisions. Genetic mosaics may also be produced in *Drosophila* by the induction with X-rays of mitotic crossing-over. If embryos or larvae heterozygous for appropriate recessive autonomous cuticular markers are irradiated, chromosome exchanges will result in the production of a clone of homozygous mutant cells. Alternatively, if both homologous chromosomes carry suitable cuticular markers, the products of a single exchange can be seen as a twin spot (figure 7.10).

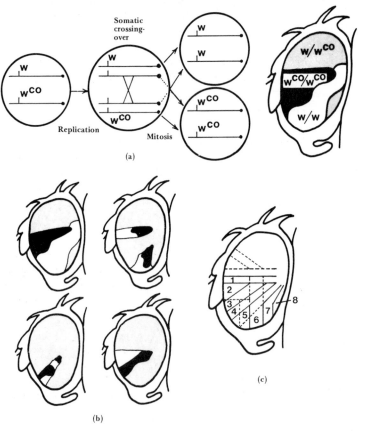

Figure 7.10   Somatic crossing-over and the production of variegated eyes in *Drosophila*. (a) somatic crossing-over giving rise to twin spots; (b) examples of variegated eyes; (c) clonal sectors identified by comparison of numerous variegated eyes. (Modified from Becker).

Becker (1957) was the first to demonstrate that mitotic recombination induced in the mid-first larval instar eye discs produced about eight sectors in the ventral half of the eye. Does the rather regular location of these sectors of the eye result from the clonal development of eight progenitor cells as discussed previously, or does the size and shape of the sector depend on the sub-division of the structure as development proceeds? Garcia-Bellido and colleagues (1973) have used a modified technique of somatic crossing-over to examine these possibilities more rigorously in the wing. They took advantage of a *Minute* (*M*) mutant which results in slow cellular growth and the formation of small bristles. The irradiation of $mwh\,M^+/mwh^+\,M$ individuals leads to the production of $mwh\,M^+/mwh\,M^+$ cells that are visibly marked and also grow faster than their neighbours. Despite this fast growth, they found that clones come to occupy only restricted areas which they termed *compartments*; a clone could come to occupy nearly the whole of an area because of its rapid cell division, but would not cross the boundary into the next compartment. In this way, the boundary lines between compartments could be seen as smooth interfaces between $mwh\,M^+$ cells and $mwh^+\,M$ cells. Within a compartment, cell lineage seems to be much less well defined, but clonal development from a small number of progenitor cells will still tend to give a fairly regular arrangement of sectors, as seen in the eye. Interestingly, the compartments do not correspond exactly with the imaginal discs of the larva. For example, the earliest division in the cells which will form the thorax and wing is into anterior and posterior compartments (figure 7.11).

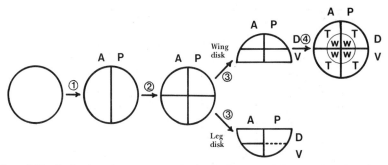

Figure 7.11 Derivation of compartments in the development of the mesothoracic cuticular structures of *Drosophila*:

A—anterior
P—posterior
D—dorsal

V—ventral
T—thorax
W—wing blade

(Redrawn from Morata and Lawrence).

These then become divided into separate thoracic and wing compartments (giving a total of four compartments) which ultimately form two imaginal discs each of two compartments. Each compartment may be further subdivided one or more times. Thus, the assignation of a cell to a particular compartment is the result of a sequence of restrictions on cell migration. The way in which this is achieved is not a question of cell lineage but of autonomy and response of the cell to position within the developing organism, and it will therefore be considered later in this chapter.

Genetic studies of determination

Grafting experiments have traditionally been used to ascertain whether a particular tissue has become determined to differentiate in a particular way, or whether it depends on the influence of surrounding tissues. Thus, a piece of tissue may be grafted to an abnormal site to see whether it retains its capacity to differentiate, and the inductive properties of certain tissues (e.g. the dorsal lip of the amphibian blastopore) can be similarly tested by examining their effects on cells that are not usually exposed to the inducer. The repertoire of such experiments can be increased in two ways: firstly, grafted tissues may be genetically distinct from the host tissues and, secondly, the production of genetic mosaics by the methods described previously can be used in place of grafting.

The grafting of genetically distinct tissues has been extensively used in *Drosophila* to establish whether a mutation acts autonomously or not. In a classical series of experiments, Beadle and Ephrussi (1937) investigated a number of metabolic mutants in the pathway of brown (ommochrome) eye pigment synthesis. They transplanted the eye imaginal discs from larvae of one genotype to larvae of a differing genotype. The donor eye disc develops at the ectopic site and, following the differentiation of adult tissues during the pupal period, can be dissected from the body of the host for examination. Typical results for two mutant alleles are shown in table 7.3. Clearly, neither the *vermilion* (*v*) nor *cinnabar* (*cn*) alleles have an eye-autonomous action since wild type host tissue obscures the mutant phenotype in transplanted eye discs. The simplest interpretation of this is that metabolic precursors of eye pigment are elaborated outside the eye discs. Mutant eye discs in wild type hosts will be provided therefore with missing metabolites. Comparison of the double mutant combinations (*v* host with *cn* donor, and *cn* host with *v* donor) indicate that the v^+ gene acts earlier in the pathway than cn^+. This has subsequently been confirmed by direct biochemical analysis.

Table 7.3 Pigment formation in grafted *Drosophila* eyes

Donor genotype	Host genotype	Type of pigment
$+/+$	$+/+$	$+$
v/v	$+/+$	$+$
cn/cn	$+/+$	$+$
$+/+$	v/v	$+$
$+/+$	cn/cn	$+$
v/v	cn/cn	$+$
cn/cn	v/v	mutant
cn/cn	cn/cn	mutant
v/v	v/v	mutant

Homoeosis and transdetermination in *Drosophila*

Imaginal discs may be transplanted into larval hosts where they will differentiate along with the host imaginal discs during the pupal period. Alternatively, discs may be transplanted into adult female hosts. In the

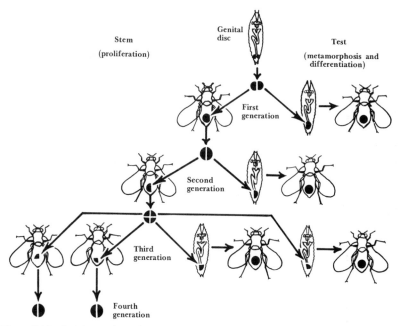

Figure 7.12 Serial transfers of imaginal discs. In each generation, disc tissue may be transferred to another adult to maintain the stem line, or implanted into a larval host which metamorphoses to reveal the current state of determination of the disc. (Redrawn from Suzuki).

haemolymph of the adult fly where there is a low titre of ecdysone, they will proliferate but will not differentiate. Morphological differentiation requires a higher ecdysone titre which can be provided by transferring the implant to a metamorphosing larva. Long-term cultures can be established by the serial transfer of discs or disc fragments from one adult generation to the next (figure 7.12). The stability of the initial determination is demonstrated by, for example, the production of anal plate after more than 70 transfer generations or approximately 1000 cell replication cycles (Hadorn, 1967). However, allotypic structures which normally arise from other discs are also produced by the proliferating disc tissue; this is the phenomenon of transdetermination first described by Hadorn. The direction of transdetermination is not random; allotypic structures appear in a certain sequence (figure 7.13) and at a certain frequency depending on the particular change in determinative state. Intermediate forms have not been described, so there seems to be some form of "switching" between the various discrete states of determination. Transdetermination occurs most frequently when the cells of the implanted disc are dividing rapidly; it has been suggested that this might dilute the level of a cytoplasmic determinant below a critical threshold.

Most of the transdetermination steps are paralleled by the transforma-

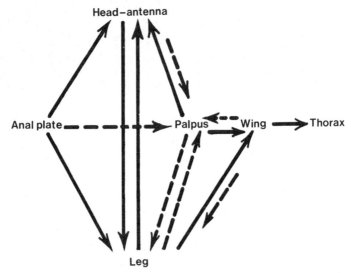

Figure 7.13 Pathways of transdetermination in *Drosophila* imaginal discs. Broken lines indicate very rare changes. (After Hadorn).

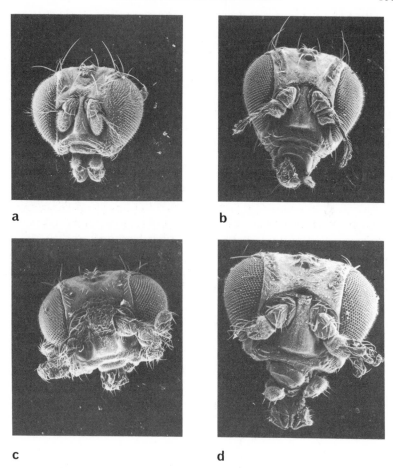

Figure 7.14   Homoetic mutants in *Drosophila*.
(a) Wild type head; (b) ss^a/ss^a, with arista transformed into leg-like structure; (c) and (d) *Antp*, showing differing extents of transformation of antenna into leg-like structures.

tions of homoeotic mutants (see chapter 6). However, the allotypic structures produced as a result of the action of homoeotic genes are entirely predictable—a particular autotypic structure is always replaced by the same allotypic structure (as in figure 7.14). A further feature of homoeosis is that adult structures are rarely entirely allotypic but contain variable amounts of autotypic and allotypic tissue.

A list of the more commonly used homoeotic mutants is provided in

Table 7.4 Homoeotic mutants in *Drosophila*

Mutant	Symbol	Chromosome	Homoeotic effect
antennapedia	*Antp*	3	antenna into leg
aristapedia	*ss*^{a}	3	arista into leg
engrailed	*en*	2	posterior wing into anterior wing
eyes-reduced	*eyr*	3	wing tissue in eye region
nasobemia	*Ns*	3	antenna into mesothoracic leg
ophthalmoptera	*opht*	3	wing tissue in eye region
polycomb	*Pc*	3	mesothoracic leg into prothoracic leg
podoptera	*Pod*	multigenic	wing into leg
proboscipedia	*pb*	3	oral lobes into tarsus-like or arista-like appendages
pointed wing	*Pw*	3	antenna into wing
wingless	*wg*	2	wing into notum

table 7.4. (The *bithorax* series is not included as this is dealt with separately in chapter 6). Mitotic recombination studies have shown that, with the exception of *wg*, these mutants are cell-autonomous. The time of action of the *antennapedia* (*Antp*) mutant has been examined in detail by Postlethwait and Schneiderman (1969). By inducing marked clones throughout larval development, they were able to show that it is only in the early third instar that such clones become restricted to either the allotypic leg structures or the autotypic antennal structures. The change in determination directed by *Antp* must occur therefore at this stage of development. Similar conclusions were reached by Grigliatti and Suzuki (1971) in temperature shift experiments with the temperature-sensitive *ss*^{a40a} mutant. In contrast, in *bx/pbx* larvae (transformation of haltere into wing), an increase in the size of the haltere disc is already apparent by the first larval instar, yet the production of *bx/bx* clones in *bx/bx*$^+$ larvae by X-ray induced mitotic recombination in the third instar will still lead to the expression of the *bx* phenotype (Morata and Garcia-Bellido, 1976). In this case therefore, it would appear that the continuous action of the *bx*$^+$ gene is required throughout much of larval development (and perhaps beyond) to ensure that development follows the normal pathway.

The parallels between homoeosis and transdetermination suggest that both processes destabilize development in analogous ways. Homoeosis is not however *in situ* transdetermination arising from increased cellular proliferation in the homoeotic disc, for the following reasons. (1) Although the allotypic structure is frequently larger than the autotypic structure, this is not always the case. For example, the *tetraltera* and *podoptera* mutants result in the transformation of wing into haltere. (2) Homoeotic transformation is restricted to the replacement of a particular autotypic structure with a particular allotypic structure, whereas the changes in transdetermination, although limited, are less restricted. (3) Homoeotic change is dependent on the presence of a particular mutant gene and the transformation that occurs is characteristic of that mutant gene.

An important feature of homoeosis first reported by Garcia-Bellido and colleagues (1973) is that the areas affected by certain homoeotic mutants correspond rather precisely with the compartments described in the previous section on cell lineage. For example, *contrabithorax* (*Cbx*) transforms the posterior mesothorax into posterior metathorax (see table 6.1); this corresponds to the compartment defined in the larva before division into separate leg and wing compartments which subsequently produce the two different imaginal discs. It is suggested, therefore, that the function of the wild type alleles at these loci is to control the development of each compartment appropriate to its position. As we have seen, the division of cells into compartments is not a single event, but a succession of sub-divisions into smaller and smaller areas. Thus, each compartment would need to be controlled by a separate gene locus. There is some evidence for this. The *bx* and *pbx* mutants transform the entire compartments corresponding to the anterior metathorax and posterior metathorax respectively. The bx^+ and pbx^+ genes could therefore control the development of these two compartments. The boundary between these two compartments is affected by the mutant *en*. This mutant leads to the transformation of posterior compartment into anterior compartment of the wing disc and to similar anterior-anterior duplications in all the thoracic discs (Garcia-Bellido, 1975). Homozygous *en/en* clones induced by mitotic recombination in the anterior wing compartment of en/en^+ flies do not cross the boundary but clones in the posterior compartment may cross the boundary and occupy sites in the anterior compartment. It would appear therefore that the en^+ gene is necessary for the establishment of the anterior-posterior boundary. Through the action of en^+, the cells of the posterior compartments are so altered that they are no longer able to mix with the anterior compartment cells (Morata and Lawrence, 1975).

What then are the physiological actions of homoeotic alleles? The capacity of *en/en* clones to cross the boundary between anterior and posterior compartments indicates that this homoeotic mutant may affect intercellular affinities, possibly by some alteration of the cell surface. By determining the cell-surface and cell-mingling properties of cells, the *en* gene is defining whether the compartment will be anterior or posterior. Homoeotic genes such as bx^+, where activity is required throughout development, may act to select and maintain a particular pathway of development. Garcia-Bellido has used the term *selector* to describe such genes. If mutation leads to the loss of the selector gene product, then all compartments in which that selector gene would act will be transformed, as in the case of *en*. Homoeotic mutants may identify therefore a series of selector gene loci that are required for the selection of different pathways of development and for the sub-division of developing structures into separate developmental compartments in relation to their positions in the organism.

Pattern formation within compartments in *Drosophila*

So far, we have discussed the effect of homoeotic alleles on determination of the entire compartment. It is also possible to study the action of homoeotic alleles on the development of the different cell types which differentiate within a compartment. For example, Stern (1968) generated a series of male flies with mosaic patches of various sizes homozygous for the *extra sex comb* (*esc*) mutant. This mutant transforms the mesothoracic legs into prothoracic legs, which in male flies possess a row of 10–13 teeth, the sex comb (figure 7.15). Two main features are seen in these flies; the mutant patches develop autonomously according to their own genotype, and they form structures appropriate to their position in the limb. Thus the state of

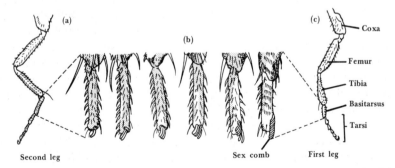

Figure 7.15 Genetic mosaics for increasing areas of the *extra sex-comb* homoetic allele, which transforms the basitarsus of the second leg so that it resembles the first leg in male *Drosophila*. The allele acts autonomously to produce large, sex-comb-like bristles, but the orientation of these depends on the size of the surrounding mosaic tissue. (After Stern and Horder).

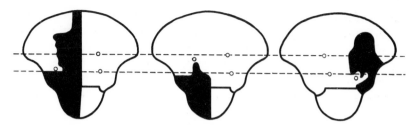

Figure 7.16   Achaete mosaics in *Drosophila*. The black areas indicate *achaete* mutant tissue, the circles the actual sites of the bristles. (Redrawn from Stern).

determination and differentiation as prothoracic leg tissue (i.e. sex comb rather than a final transverse basitarsal row of bristles) does not depend on their surroundings, whereas their differentiation into a particular structure within the range of possibilities for that limb (i.e. sex comb rather than any other type of bristle) depends entirely on the surrounding tissue which supplies some kind of positional information.

In a classical experiment, Stern (1968) generated flies mosaic for the *achaete* (*ac*) mutant which leads to the lack of thoracic bristles (figure 7.16). Bristles fail to develop at their normal sites if these are occupied by mutant tissue. If there is normal tissue nearby, a bristle may develop there instead, in a position where it is not normally found. This suggested to Stern that there was some kind of "prepattern" or positional information which determines where the bristles should develop. If the tissue is genetically not competent to respond, the bristle could develop nearby in response to the prepattern. It would not normally develop there, so that a bristle as it develops must abolish the prepattern or otherwise inhibit the formation of another bristle nearby. This concept of prepattern is further supported by the behaviour of the fourth chromosome *eyeless-dominant* (*ey*^{D}) mutant (Stern and Tokunaga, 1967). In addition to its effect on the development of the compound eye, this mutant produces in male flies three or more times the normal number of sex comb teeth. In order to generate cuticular mosaics, Stern and Tokunaga recovered a third chromosome with inserted *ey*^{D} that also carried a duplication of *y*$^+$ at its tip. They then irradiated male larvae of this genotype which also carried *y* on the X-chromosome. Mitotic recombinants induced in the region between *ey*^{D} and the centromere result in the loss of both *y*$^+$ and *ey*^{D} in the daughter cell, and is detectable as a yellow patch. A number of mosaics with *y* sectors in the sex comb region were recovered, and in every case multiple sex combs were formed even though the cells themselves did not carry the *ey*^{D}

gene. This is therefore a case of non-autonomy where the prepattern established by the surrounding $ey^D/+$ tissue is interpreted by the $+/+$ tissue to give the ey^D phenotype.

The nature of this positional information in development is unknown. Suggestions include a chemical or physical gradient, or a wave form produced by two signals emitted at different relative amplitudes. Alternatively, the information could be provided by means of the specific interactions of embryonic induction. The experiments described here do not resolve this issue, and further discussion of this problem is outside the scope of this book. Nevertheless, the analysis of genetic variants which do not respond to position within a compartment might provide valuable insights into this problem.

Applicability of the compartment model to vertebrates
Since cell lineage is considerably more complex in vertebrates, homoeotic mutations may be hard to recognize as such, and given the importance of intercellular interactions, such as embryonic induction, in bringing about the physical organization of the embryo, they might well be lethal. The *ametapodia* mutant in the fowl which causes the wing to develop approximately as a leg is, however, one example of homoeosis in vertebrates (Cole, 1967). Experimentally, mesoderm from the hind-limb bud when transplanted underneath fore-limb apical ectoderm, develops autonomously into hind-limb structures in its new position, and also induces the fore-limb ectoderm to form hind-limb structures such as scales and claws. Thus *ametapodia* is probably expressed initially in the wing mesoderm, but its effect is also seen in ectoderm cells (of a completely different cell-lineage). From this example, it would appear that the restriction of homoeotic gene effects to a single compartment as seen in *Drosophila* does not apply directly to vertebrates. Although the same elements of cell lineage and determination are involved, cell movement and cell mingling are considerably more extensive, and determination appears to depend to a greater extent on embryonic induction between cell lineages. For other genetic studies of induction and limb development, see Zwilling (1956a,b,c).

Conclusion
In mammalian chimaeras, cell lineage relationships for most tissues can be established for only the final few cell divisions. This is because clonal development is usually disrupted by extensive cell mingling and cell

migrations; it is unlikely therefore that cell lineage plays a direct role in the processes of determination and differentiation. The disruption of coherent clonal development also means that, with the few exceptions where precise cell lineages can be traced (e.g. the photoreceptor cell layer of chimaeric mouse retina), the validity of estimates of tissue precursor cell number based on the spatial arrangements of genetically marked clones is questionable. An alternative approach is to consider the relative content of the two cell types in a chimaeric tissue. However, if this is done, then the competitive interaction and the differential viabilities of the two cell types must be quantified; the difficulty here is that the interactions of cells from different mouse strains are tissue specific and may even vary at different developmental stages.

In contrast, development in *Drosophila* does not involve extensive cell movement and the position of cells in the blastoderm layer of the embryo plays a significant role in deciding fate; destruction of precise areas of the blastoderm by a variety of techniques (see page 47) leads to predictable lesions in the resulting individuals. Cytoplasmic determinants present in the cortex of the egg that direct the subsequent development of the cells that come to contain them could account for these observations. However, cell lineage cannot be the only factor since the progeny of a single cell (i.e. a M^+/M^+ cell produced by mitotic recombination in otherwise M/M^+ tissue) can come to occupy most of a compartment. Position within the blastoderm could serve to direct cells into particular compartments, but the final allocation of cells to a particular pathway may not occur until later in development. Within a compartment, development would then depend on position and cellular interactions.

With successive sub-division of compartments, the selection of appropriate pathways of development depends on the activity of selector genes. Homoeotic development arises from mutations in selector genes diverting development along alternative pathways. The bx^+ locus, for example, is known to be required throughout development for the maintenance of normal development of the anterior metathoracic compartment, but the precise mechanism of gene action is unknown. The bx gene complex contains a number of loci important for the normal development of the thorax, and gene expression in heterozygotes for different mutant alleles depends on the chromosomal location of the respective mutants—this topic was considered in greater detail in chapter 6.

CHAPTER EIGHT

INTERCELLULAR INTERACTIONS IN DEVELOPMENT

In development, multiple phenotypic effects or pleiotropy arising as a result of a single mutation may be produced in a number of distinctive ways. Small deletions may involve the loss of a number of gene loci; each deleted locus may have a consequent effect on development. Such a situation is encountered with the radiation-induced *lethal-albino* mutants (Gluecksohn-Waelsch, 1979) where the effects on pigmentation (when heterozygous with a non-lethal *albino* allele), embryonic development and plasma protein levels are the result of the loss of a small chromosome segment surrounding the *albino* gene. The search for a common underlying factor in the production of the pleiotropy in such cases would clearly be unproductive. There are however many examples where the mutation is confined to a single gene, yet results in multiple developmental effects. In such cases, the primary action of the mutant gene must be either to interfere with a fundamental process in cellular homeostasis common to many tissues (see also p. 110) or to modify or abolish interactions between cells or tissues.

The mottled mutants in the mouse

Six alleles have been described at the X-linked *mottled* (*Mo*) locus in the mouse (table 8.1). Female heterozygotes for any allele (*Mo*/+) characteristically show pigment variegation in the coat (see figure 5.3) as a result of inactivation of either the X-chromosome carrying the normal (+) allele or the mutant (*Mo*) allele in different pigment cell clones. Mutant males (*Mo*/−) show a complex pattern of pleiotropic effects. These include a severe pigment deficiency throughout the coat (apparent in mutant males that survive beyond birth), a hair structure defect especially apparent in

Table 8.1 Phenotypic characteristics of mice hemizygous for alleles at the X-linked mottled locus in the mouse

Allele	Hemizygous phenotype
Blotchy (Mo^{blo})	Pale fur, curly whiskers. Viable and fertile
Brindled (Mo^{br})	Pale fur, curly whiskers. Dies at about 14 days after birth
Dappled (Mo^{dp})	Pale fur, curly whiskers, skeletal defects. Usually dies at birth
Mottled (Mo)	Lethal *in utero*
Viable-brindled (Mo^{vbr})	Pale fur, curly whiskers. Reduced viability and sterile
Tortoiseshell (Mo^{To})	Usually dies *in utero*

the whiskers (figure 8.1), bone deformities (in some alleles), arterial aneurisms in older surviving mutant males, neurological disturbances and pre- or post-natal death in many alleles (table 8.1).

In a study of the neurological defect in this disorder in $Mo^{br}/-$ mice, Hunt and Johnson (1972) showed that brain levels of the neurotransmitter noradrenaline were reduced in $Mo^{br}/-$ mice as a result of the depressed activity of dopamine β-monooxygenase. When this was considered alongside the other effects of this locus, a common factor of copper metabolism emerged; tyrosinase and dopamine β-monooxygenase are both copper-metalloproteins, and the synthesis of keratin (in hair), collagen (in the formation of the bone) and elastin (in arterial walls) are known to contain copper-dependent steps; a defect in copper metabolism was subsequently looked for and found (Hunt, 1974). The pleiotropic effects of the alleles at

Figure 8.1 Hemizygous Mo^{br} mutant male mouse. Note the lack of hair pigment and the abnormal texture of the coat. The animal also has curly whiskers.

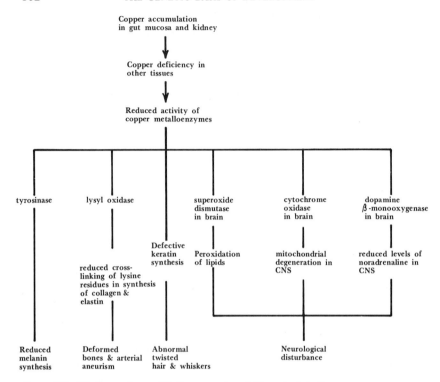

Figure 8.2   "Pedigree of causes" in the expression of *Mo* mutants in the mouse.

this locus can now be understood in terms of the severity of their effect on copper homeostasis (figure 8.2). In mutant mice, copper accumulates in the gut mucosa and in the kidney, and is deficient elsewhere (particularly in the liver and brain). The disorder, although expressed in many tissues as copper deficiency, arises therefore from copper retention in kidney and gut tissue (Port and Hunt, 1979).

The *dominant-spotting* and *steel* loci in the mouse

Mice heterozygous for the *dominant-spotting* (*W*) gene generally have a well-defined white spot in the belly region, with a very variable amount of white hair in the dorsal coat. Homozygotes for the viable allele W^v are completely white (with black eyes); the white spot includes the entire coat in these mice. Most *W/W* mice die at birth with a severe macrocytic anaemia, and anaemia is also seen in surviving W/W^v and W^v/W^v mice. In addition, W^v/W^v mice have abnormalities in the cochlea of the inner ear (Deol, 1970) and are usually sterile. The common factor is that, in each case, the affected tissue is populated by cells arising during embryogenesis at a site distant from their adult location. Mutant gonads are drastically reduced in germ cells (Coulombre and Russell, 1954), the unpigmented "spot" regions are devoid of identifiable melanocytes (Silvers, 1956), and the anaemia is of the hypoblastic type (Russell, 1970), originating from an initially-depleted colonizing cell population.

The simplest explanation for the effects of W is that it affects the ability of these cells to interact with and populate their target sites; the defect is autonomously expressed in the migratory cells. The lesion in the inner ear may also have a similar aetiology since the development of the disorder involves the acoustic ganglion, a component of which is derived from the neural crest, the site of origin of the melanoblasts.

The alleles at the *steel* (Sl) locus also affect melanogenesis, erythro-poiesis, and germ cell development. However, in contrast to W, the defect in Sl adversely affects the ability of the target tissue to support the subsequent development of the migratory cells; the defect in Sl is non-autonomously expressed in these cells. For example, when epidermis from 13 day-old $+/+$ embryos was combined with Sl/Sl^d dermis, and trans-planted into the coelom of a chick embryo (a convenient supportive system for subsequent development of the recombined skin), all grafts failed to display pigment in the dermal region yet produced pigmented hairs (of epidermal origin). In reciprocal combinations, the $+/+$ dermis of the graft was pigmented but the hairs were uniformly lacking in pigment (Mayer, 1973). Likewise, when marrow cells from Sl/Sl^d animals were implanted into heavily irradiated normal mice, they were able to colonize the spleen, whereas $+/+$ marrow cells failed to proliferate and differentiate normally in irradiated Sl/Sl^d mice (McCulloch and colleagues, 1965).

Genetic basis of endocrine systems

As previously discussed (page 99), a single hormone will elicit diverse responses in different target tissues; the precise response depends on the developmental programming of the responding cells. Since each step from the production of the hormone (whether it is a polypeptide and hence a direct translation product of a gene, or a biologically-active molecule such as a steroid that is elaborated by the enzymes of a biosynthetic pathway), its controlled release from the cell, its transport in the body fluids, its interaction with the target organ, to its final degradation and excretion must be gene-controlled, it follows that mutations affecting each step will occur. In many cases, such mutations are seen as a series of pleiotropic effects. For example, the *Tfm* mutation prevents the recognition of testosterone by target cells. The resulting phenotypic effects include rudimentary testes, female genitalia and mammary gland development in genotypic males carrying *Tfm* on the X-chromosome (see chapter 9). Pituitary dwarfism in the mouse may be caused by two unlinked genes, *dw* and *df*. In both cases reduced growth is associated with sterility and absence of the X-zone of the adrenal cortex; the developmental effects of both disorders are consistent with deficiencies of growth hormone and prolactin (Shire and Hambly, 1973). Insulin diabetes occurs in mice homozygous for the db^{2J} gene, although the expression of the disease depends on particular combinations of modifier genes present in different strains; on the C57BL/KsJ background, the diabetes is severe and marked hyperglycaemia and degeneration of islet cells is seen whereas, on the C57BL/6J background, the diabetes is mild, with elevated insulin, transient hyperglycaemia and hypertrophy of the islets (Hummel and colleagues, 1972).

In the above examples, the examination of genetic variation affecting endocrine systems has successfully led to the identification of the diverse effects of a single hormone. Disease states may also vary according to interactions of a mutant gene with its genetic background, as seen for insulin diabetes in the mouse (see Shire, 1976, for a more detailed discussion). The genetic makeup of a strain can also influence the way a particular homeostatic system is assembled. In the production of adrenaline by the adrenal medulla, the activity of the enzyme phenyl-ethanolamine N-methyltransferase responds to a number of control systems. These control systems may not however be the same in all strains of mice (Ciaranello, 1978). In the DBA/2 strain, enzyme activity can be modified by the secretion of adrenal glucocorticoids and directly by the nervous system via the splanchnic nerve, whereas in the CBA strain, only glucocorticoids are effective, and in the C57BL/Ka strain the activity of the enzyme is under the direct control of adrenocorticotrophin secreted by the pituitary.

Mutations affecting metamorphosis in *Drosophila*

In *Drosophila*, the presence of moulting hormone, ecdysone, is essential for the development of adult structures. Even imaginal discs cultured *in vitro* in a chemically-defined medium (Mandaron, 1971) will undergo metamorphosis on the addition of ecdysone. As a result therefore of this single hormonal trigger, diverse developmental programmes in the different imaginal discs and involving the differential expression of further gene loci are initiated, with the result that large qualitative differences are produced between tissues which were previously sufficiently similar for transdetermination to occur between them (see chapter 7).

Mutant alleles at several different loci are known which lead in different ways to developmental arrest at pupation. Most larvae homozygous for the *lethal-giant-larvae* (*lgl*) mutant on chromosome 2 fail to pupate and instead become very large. The ring gland of such larvae is small (Aggarwal and King, 1969), with pronounced deficiencies of smooth endoplasmic reticulum in prothoracic gland cells, the source cells for ecdysone, and the implantation of wild type ring glands will induce puparium formation. The condition however cannot be due to a simple failure in ecdysone biosynthesis or release since the third instar larvae have already undergone two ecdysone-requiring larval moults. The primary defect appears to be in the brain, where the presumptive adult optic nerves proliferate and become invasive (Gateff and Schneiderman, 1974). This could cause either a deficiency of the brain hormone which normally stimulates ecdysone secretion from the prothoracic region of the ring gland, or else a failure of the brain to carry out its normal inhibition of the copora allata during the third instar, with the result that elevated juvenile hormone may block ecdysone production or release.

A defect in hormone secretion of the kind seen in *lgl* will be non-autonomously expressed in target tissues of genetic mosaic individuals, since all the tissues will be subject to a similar stimulus (whether normal or deficient depending on the genotype of the tissue expressing the primary defect). A defect in response to hormone secretion, on the other hand, will be autonomous because only the abnormal cells will fail to respond. Kiss and colleagues (1976) used this approach to test a series of six recessive X-linked alleles induced by the mutagen, ethyl methanesulphonate, which are lethal at the end of the third larval instar. Ring-X chromosome elimination at cleavage (with consequent loss of the wild type allele) was used to generate gynandromorphs with the mutant allele on the retained rod-X. In each case, XO regions of the larva consistently failed to pupate whereas XX regions produced tanned cuticle characteristic of puparium formation. The autonomous expression of the mutants, together with the normal development of mutant larvae through two larval moults, suggests that the production and release of ecdysone is not defective, and that the failure to respond is specific to the larval/pupal moult.

Developmental physiology of the moss, *Physcomitrella patens*

In plants, there is a smaller distinction between physiology and development than in animals. This reflects the fact that most plants grow throughout their lifetimes, and the growth is accompanied by the continuous differentiation of new structures. They also show marked growth responses to their physical environment, especially light and gravity.

Cove and his colleagues (Ashton *et al.*, 1979) have chosen a moss, *Physcomitrella patens*, for the genetic analysis of plant physiology and development. The advantages of this plant are that it is sufficiently complex to show many features of higher plant development but, at the same time, it can be grown on a defined culture medium and handled by many of the techniques of microbial genetics. The gametophyte stage of development can be conveniently divided into a number of phases. Following spore germination, branching filaments of tubular cells with perpendicular cross-walls (chloronemata) grow out. After about 7 days caulonemata are produced, composed of larger tubular cells with oblique cross-walls. Caulonemata subsequently divide to form side branches, the majority of which become secondary chloronemata. From 10–11 days after germination buds develop on a few of the caulonemal side branches. These subsequently differentiate into gametophores with a stem and leaves.

Following treatment of spores with mutagens, abnormal clones can be detected either by their morphological appearance or by some screening procedure such as ability to grow on normally toxic concentrations of auxin or cytokinin. Although the morphological abnormalities of most mutants lead to sterility, a degree of genetic analysis can still be achieved with somatic hybrids following protoplast fusion with polyethylene glycol. Such hybrids constitute a test for genetic complementation between two different mutants (i.e. $a +/+ b$ will, in general, have a wild type phenotype if a and b are recessive mutant alleles at different loci, but a mutant phenotype if they are alleles at the same locus).

In this way an array of mutants of different types have been isolated, including some with defective biosynthesis of auxins and cytokinins; their analysis indicates that auxin and cytokinin play interdependent roles in the regulation of gametophyte development. The model presented by Ashton and co-workers to account for the effects of auxin and cytokinin on wild type and mutant development is shown in figure 8.3. Mutant clones blocked in auxin synthesis produce more chloronemata but fewer gametophores. These mutants are also resistant to the action of cytokinin unless pretreated with auxin, suggesting that the response to cytokinin is

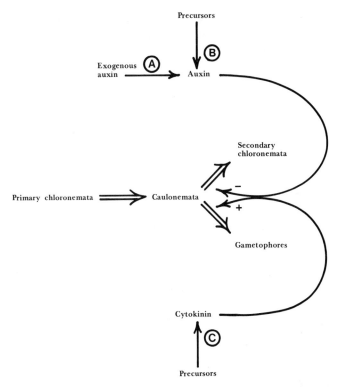

Figure 8.3 A genetic and physiological model of gametophore development in the moss *Physcomitrella patens*. Mutant clones have been isolated which are defective in (A) the uptake of exogenous auxins, (B) the biosynthesis of endogenous auxins, and (C) the biosynthesis of cytokinins. (After Ashton and colleagues).

auxin-dependent. Low levels of auxin are probably required for caulonemal development, as some auxin-deficient mutants will only make caulonemata when supplied with exogenous auxin. Cytokinin-deficient mutants produce no gametophores but overproduce secondary chloronemata, and can be repaired with exogenous cytokinin. In the absence of exogenous cytokinin, exogenous auxin has little effect, suggesting that both these substances are necessary simultaneously for the inhibition of secondary chloronemal development and stimulation of gametophore development. It is also possible to obtain mutant clones which are defective in their response to light of differing wavelengths and polarization (Cove and colleagues, 1978). In the same way that genetic analysis has permitted the dissection of the actions of auxin and cytokinin, these mutants are providing a tool for investigating phototropic responses. For example, even though protonema from the germinating spore are unicellular and gametophore development is a multicellular process, the light responses of both tissues are affected in some mutant clones; their phototropic responses must therefore include one (or more) common components.

t-alleles in the early development of the mouse

An allele known as brachyury ($=$ *short tail* or *T*), which caused the absence of the terminal tail vertebrae, was discovered in the mouse during an irradiation experiment carried out in Paris by Dobrovolskaia-Zavadskaia in 1924. She found that T/T embryos died *in utero*, but when outcrossed to other stocks, some of the progeny were tail-less rather than short-tailed, completely lacking in caudal vertebrae. The factors responsible for this effect were identified as *t* alleles; tail-less mice were genetically T/t. A number of different *t* alleles have subsequently been identified. When homozygous, they frequently cause embryonic lethality but are recessive to wild type and only affect the tail when heterozygous with *T* (table 8.2).

The *t* alleles show a number of unusual features (see review by Bennett, 1975). They are genetically unstable and give rise to new *t* alleles at substantial rates. They may also cause sterility or show segregation distortion (where the two alleles in the $t/+$ heterozygote are not equally represented in the gametes) in the male, and cross-over suppression over the region of chromosome 17 adjacent to their location.

Formal genetic analysis can be carried out with crosses in which the segregation of *t* alleles is followed with the aid of a closely-linked marker,

Table 8.2　Developmental effects of T and t mutants

Genotype	Phenotype
$T/+$	short tail
T/T	embryonic lethal
$t/+$	normal tail
t/T	tail-less
t^x/t^x	embryonic lethal, semilethal or viable depending on allele
t^x/t^y	embryonic lethal, semilethal or viable depending on allelic combination

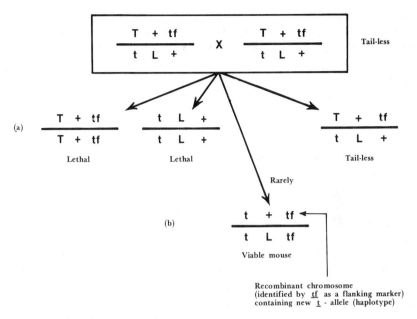

Figure 8.4 (a) The maintenance of *t*-alleles in a balanced lethal system which permits genetic analysis of the progeny in relation to the flanking marker, tufted (*tf*).

(b) An explanation for the origin of a new viable *t*-allele (haplotype) by recombination. *L* is very closely linked to *tf*; *t* + is the new *t*-allele (haplotype).

tf (tufted) (figure 8.4). With *t* alleles that are embryonic lethal when homozygous (e.g. t^6), progeny are occasionally produced which carry a *t* allele which is no longer lethal, although the effect on the tail with *T* is still present; the production of these new *t* alleles is invariably associated with recombination between *t* and *tf*. An interpretation of this phenomenon is given in figure 8.4. The segregation of the lethality effect from the tail-length defect suggests that two separate loci are involved, which can be designated *T* and *L* (for lethality). The new allele is a recombinant between the existing alleles at these two loci. The *t* locus is therefore a complex locus (see chapter 6) made up of several genes, and it is more appropriate to refer to the *t*-alleles as *haplotypes*. This term indicates a particular combination of alleles in the *t* region of one chromosome. So far three loci have been identified by recombination experiments of this kind, *T* affecting tail-length (mutant also in brachyury-*T*), *A* distorting segregation and *LS* with effects on both lethality and male sterility. This last locus could be separable into two by recombination, but this has yet to be achieved. The

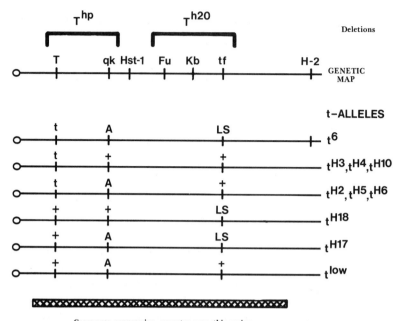

Figure 8.5 *t*-alleles interpreted as haplotypes of a complex locus. Three loci (*T*, *A* and *LS*) are identified by recombination. *t*⁶ gave rise to the other alleles in this diagram. (Modified from Lyon and colleagues).

genetic map for these three genes (Lyon and Mason, 1977; Lyon and colleagues, 1979) is shown in figure 8.5. The *t*ʰ²⁰ and *T*ʰᵖ mutants are thought to be deletions and their predicted positions are also indicated.

Turning now to a more detailed consideration of the effects of *t* on embryonic development, three groups can be identified on the basis of their viability when homozygous-viable, semi-viable and lethal. The viable alleles (*t*ᵛ) are presumably haplotypes that are wild type for the *LS* factor. In the semi-lethal group (*t*ˢ), a variable proportion of homozygotes carrying these haplotypes, ranging from 49–98%, die *in utero*; any survivors are morphologically normal. The lethal haplotypes (*t*ˡ) fall into six groups by complementation tests (i.e. if the heterozygote for two lethal haplotypes is also lethal, then they are assigned to the same complementation group, whereas, if the heterozygote survives and is morphologically normal, then they are members of different groups). These groups are listed in table 8.3. Complementation between members of different groups is not

Table 8.3 Division of some t-alleles into groups on the basis of lethality complementation properties

Viability of homozygote	Complementation group	Alleles
Lethal (t^l)	t^0	$t^0, t^1, t^6, t^{30}, t^{h16}, t^{h7}, t^{h13}, t^{h18}$
	t^9	$t^4, t^9, t^{w18}, t^{w30}, t^{w52}$
	t^{12}	t^{12}, t^{w32}
	t^w	$t^{w5}, t^{w6}, t^{w10}, t^{w11}, t^{w13}, t^{w14}, t^{w15}, t^{w16}$
	t^{w1}	$t^{w1}, t^{w3}, t^{w12}, t^{w20}, t^{w21}, t^{w71}, t^{w72}$
	t^{w73}	t^{w73}
Semi-lethal (t^s)		$t^{w2}, t^{w8}, t^{w36}, t^{w49}$
Viable (t^v)		66 alleles

however always complete and a variable proportion of heterozygotes die *in utero*.

The simplest interpretation of the six complementation groups is that there are six closely-linked loci within the *LS* region; a haplotype may be mutant at one or more of these loci. Complementation in the compound heterozygote (t^x/t^y) will occur if the two haplotypes are mutant at different loci within the region. If two haplotypes are mutant for overlapping sets of loci, then they will not complement either with each other, or with haplotypes mutant for any of the same loci (table 8.4). This may account for some of the difficulty of assigning certain haplotypes to complementation groups. In addition, not all mutant alleles at any one locus are necessarily identical and they could therefore differ in their biochemical and complementary properties; intra-genic complementation could occur to further complicate the analysis.

This model can also account for the occasional irreversible production of a new but complementary lethal haplotype from an existing one (Klein

Table 8.4 Complementation between hypothetical haplotypes where m denotes a mutant locus. t^a and t^b will complement each other, as will the non-overlapping t^c and t^d. However, t^e which overlaps t^b, t^c and t^d would only complement with t^a.

	Locus					Haplotype
1	2	3	4	5	6	
m	+	+	+	+	+	t^a
+	m	+	+	+	+	t^b
m	m	+	+	+	+	t^c
+	+	m	+	+	+	t^d
+	m	m	+	+	+	t^e

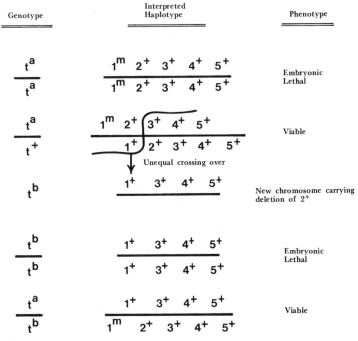

Fig. 8.6 Origin of a new lethal haplotype by unequal crossing over between t^a and t^+. m—mutant allele; $+$—wild-type allele; 1–5—complementation groups. t^b is the new chromosomal haplotype; t^b/t^b is lethal because of the deletion but t^b/t^a shows complementation and is viable because a copy of each wild-type allele is present. (Modified from Klein and Hammerberg).

and Hammerberg, 1977) by unequal crossing-over with the wild type haplotype within the *LS* region (figure 8.6). The unequal crossing-over would give rise to a new haplotype (t^b from t^a) which carried a deletion at one of the other loci. This haplotype would therefore be homozygous lethal, but would complement the first mutant haplotype defective only in the original mutant locus. Since t^b carried a deletion, it could not give rise to t^a. Despite the success of this model in explaining some of the observations, it should be stressed that other explanations are also possible.

Developmentally, the interest in *t*-haplotypes lies in the discovery that the members of each complementation group have characteristic effects on embryonic development, both in the time of death and the tissue which is most affected (table 8.5). In other words, the proposed six loci, and also

Table 8.5 Lethal development defects in homozygotes for various haplotypes in the t-region

Complementation group	First time of action (days gestation)	Affected tissue	Type of defect
t^{12}	1	Morula	Failure to form blastocyst
t^{w73}	3	Trophectoderm	Failure to implant
t^0	5	Inner cell mass	Failure of ectoderm to differentiate into embryonic and extraembryonic
t^{w5}	6	Embryonic ectoderm	Degeneration
t^9	7	Primitive streak	Uncontrolled growth
t^{w1}	9	Neural tube	Degeneration of ventral tissue
t^{w2}	14	Forebrain	Failure of eye and ear development
T	8	Notochord or primitive streak	Irregular neural tube and notochord

the t^s locus, each seem to have a distinct effect on development. The later-acting lethals affect a smaller and smaller region of the embryo, with effects restricted to the development of the ectoderm and some of its derivatives (figure 8.7). The T mutant, which is fairly close to the LS region (about 7 centiMorgans) also has effects on embryonic development, leading to the short-tail in the heterozygote described earlier. In the lethal homozygote, the entire notochord and neural tube is irregular; the primary defect is probably in the caudal notochord or in the primitive streak tissue which gives rise to it. The effect in the heterozygote is accentuated to tail-lessness if the second allele is a t-haplotype mutant in the T-region; the t-haplotypes probably produce a slightly altered gene product from the T locus.

Other T alleles (e.g. T^{hp} and T^{Orl}) appear to be deletions extending some way on either side of the T locus (figure 8.5) and so show several of the mutant effects. Complementation studies with these alleles reveal that the t^{w73} haplotype, the sole member of one of the complementation groups, carries a mutant allele (designated p) to the left of the T locus which is responsible for the defect in trophectoderm leading to failure at implantation. This complementation group is therefore not in the LS region. The mutant *knobbed* (Kb) when homozygous has an effect on development similar to the effect of t^{w5} and is very near the LS region. Figure 8.8 gives a

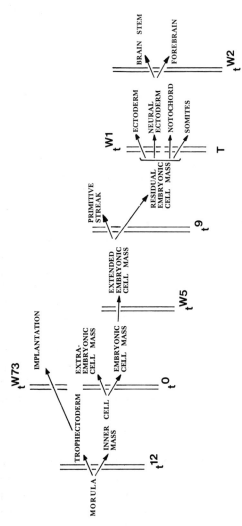

Figure 8.7 *t*-alleles in the developmental pathway of the mouse. (Modified from Bennett and colleagues).

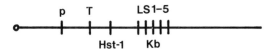

Figure 8.8 A possible map of loci in the t-region of chromosome 17 which affect embryonic development in the mouse. *Knobbed* (Kb) is probably homologous to t^{w5}; Hst-1 may also be related to the t-alleles.

provisional map of the loci affecting embryonic development (L1–L5 represent the five complementation groups arranged between kb and tf).

How do these loci bring about their effects on development? A cell surface antigen F9 was first detected on teratocarcinoma embryonal carcinoma cells (see page 25) and subsequently shown to be present on early cleavage stage embryos and on sperm. By absorbing anti-F9 antisera with sperm from different sources, Artzt and colleagues (1974) were able to show that there is less F9 antigen on sperm from male mice carrying the t^{12}-haplotype. The gene product of $+^{t^{12}}$ and the F9 antigen may therefore be identical. More recently (Shur and Bennett, 1979) elevated activities of cell-surface glycosyltransferases have been found on sperm from mice segregating for a variety of t-haplotypes, and from mice heterozygous for T^{hp}. To account for these observations, Shur and Bennett suggest that this increased activity arises from reduced inhibition of the enzyme on t and T^{hp} sperm, and that this increased activity may be at least partly responsible for the increased fertilizing ability of some t sperm (i.e. for the segregation distortion). In addition, a major cell-surface protein has been identified by two-dimensional electrophoresis of testicular proteins (Silver and colleagues, 1979) that is not produced by the T^{hp} deleted chromosome. An altered form of the protein is found with other T and t-haplotypes, either extracted from the testis or translated from testicular cytoplasmic RNA *in vitro*. It is likely therefore that this protein is specified by the T locus, and may identify the glycosyltransferase inhibitor (Silver and colleagues, 1979).

The precise relationship between these observations on sperm cell-surface proteins and the developmental arrest seen in t^l-haplotypes has yet to be determined. The defects may however arise from defective cellular interactions mediated by abnormal activities of glycosyltransferases. During chick gastrulation, migrating cells show elevated glycosyltransferase activity (Shur, 1977), and active surface transferase has also been demonstrated between notochord and somites, between the optic cup and skin ectoderm, and between medullary plate and presomite mesoderm.

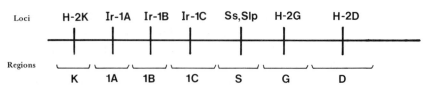

Figure 8.9   Genetic map of the H-2 complex.

If the $+^{LS}$ locus codes for a specific inhibitor of this activity in mouse embryos, the failure of this inhibition in t mutants at different developmental stages may account for the stage-specific and tissue-specific developmental arrest seen in members of the different t^1 complementation groups.

The major histocompatibility complex in the mouse

Some 13.5 cM from the T locus on chromosome 17 of the mouse is the major histocompatibility complex or H-2 complex. This complex comprises a number of loci important in graft rejection between genetically dissimilar individuals and in mounting an immune response (see Klein, 1975). It is functionally equivalent to the H-LA complex in man. At either extreme of the H-2 complex are the H-2D and H-2K loci (figure 8.9) which each specify a cell-surface glycoprotein; it is these proteins that are recognized on the donor tissue as foreign cell-surface antigens by the host immune system. A large number of different alleles at each locus has been described; each allele specifies a number of antigenic specificities which reflect different amino acid substitutions in the glycoprotein gene product. These antigenic specificities fall into two groups, the public antigens that are shared by a number of different alleles, and the private antigens that are restricted to a single allele. Each allele specifies a number of public antigens and one private antigen. A particular combination of H-2D and H-2K alleles carried on one chromosome can be characterized therefore by their private antigens; in parallel with the t complex, such genic combinations are referred to as haplotypes. However, unlike the t antigens which are restricted to specific cells, the H-2 antigens are present on most tissues except embryonic cells.

The Ia cell-surface antigens, a group of antigens restricted to lymphocytes, spermatocytes, epidermal cells and macrophages are also specified by genes contained within the H-2 complex, and a further cell-surface antigen found on thymus-derived lymphocytes is coded by the *Tla* gene located outside the H-2 complex but only 1.5 cM from H-2D. The H-2 complex also contains a number of loci (the immune-response or *Ir* genes) defined by their effect on the immune response to particular antigenic synthetic branched polypeptides. Finally, the S region codes for a serum component and the G region carries an additional histocompatibility gene, H-2G.

Do the H-2 antigens play a role in cell recognition during development? Bodmer (1972) speculated that the cell-surface proteins necessary for such interactions are specified by loci contained within the H-2 complex but distinct from H-2D or H-2K. The function of the H-2D and H-2K antigens would be to carry out some general effector function in cell recognition, thus accounting for their presence in practically all tissues. In certain respects, therefore, they would be similar to the C regions of the immunoglobulin light and heavy chains. There is, however, no direct evidence in support of this interpretation and it is generally thought that the H-2 antigens do not play a role in development.

The expression of H-2 antigens is however under developmental control. Whereas most cells acquire H-2 antigens during foetal development, their appearance on erythrocytes does not occur until around birth. All inbred mouse strains examined exhibit one of two phenotypes; H-2 antigens are acquired either just before birth (early expression) or about 3 days after birth (late expression). A cis-acting temporal gene designated *Intrinsic* (*Int*), closely-linked to the H-2D locus but outside the H-2 complex, is responsible for this difference in some strains (Boubelik and colleagues, 1975). In other strains, two unlinked loci *Tem* and *Rec* are also involved, and the temporal phenotype depends on complex interactions between *Int* and the alleles at these two loci.

Conclusions

This chapter has dealt with a number of examples of cellular interaction in development. As we have seen, pleiotropic gene action can arise from mutants affecting such interactions, either by interfering with the production or reception of a hormonal signal, or by modifying the processes of intercellular contact and recognition. With the demonstration that cell-surface antigens are specified by genes contained within the *t* complex, the stage-specific arrest of development characteristic of the different t^1 complementation groups may be understood as a failure of the normal expression of cell-surface receptors during development, with the consequent failure of particular embryonic tissues. Interaction between opposing cell surfaces could also account for the failure of *W* mutant melanoblasts to differentiate at target sites, and for the inability of *Sl* tissues to support such development. Many other developmental mutants may also depend on defective intercellular contact and recognition. For example, the establishment of compartments in imaginal disc development in *Drosophila* and the alteration of their developmental potential by the action of homoeotic genes may be achieved through the expression of cell-surface receptors. The application of techniques similar to those used in the study of the *t* mutants to these developmental systems may now be appropriate.

CHAPTER NINE

SEXUAL DIFFERENTIATION IN MAMMALS

Sexual differentiation is the process whereby male and female organisms come to possess their characteristic and different phenotypes. Apart from its intrinsic fascination, it has been widely regarded as a model for differentiation. The amplification of a small initial chromosomal difference into a larger phenotypic difference in a particular organ between males and females is analogous to the amplification of a small cytoplasmic difference between different regions of the same egg into phenotypic differences between organs within the body. The process is described here to show also how the mechanisms described separately in the previous chapters (at different levels from the gene to the whole organism) are related to each other in an integrated system and how they may be analysed by genetic techniques.

The strategy of mammalian sexual differentiation

Sexual differentiation has been particularly well studied in *Drosophila* (some aspects have been described in earlier chapters) and in mammals. In both cases, the determination is by the XX/XY chromosomal mechanism, establishing the sex of the individual at fertilization. Given this fact there are two alternative strategies for coordinating the changes which take place in different parts of the body during the process of sexual differentiation. Either the chromosomes can act autonomously within each of the cells or organs of the body, or they may act specifically only in one organ, which then establishes the sexual differentiation of the organs in a hierarchical system. Of course, there could be a combination of both strategies in the real organism! As we have seen (p. 144), XX/XO gynandromorphs in *Drosophila* develop a mosaic phenotype, so that an

autonomous system is operating. By contrast, mammalian sex chromosome chimaeras or mosaics (especially well studied in mice and men
respectively) usually develop into one sex or the other, but only occasionally into an intersex animal (McLaren, 1976). The intersex in these cases is
an animal where the external genital organs may be intermediate in their
sexual differentiation irrespective of their own particular chromosomal
sex, rather than a mosaic of two distinct types, and where both male and
female duct systems may be present in the same animal. The only organ
which is seen to develop a mosaic phenotype is the gonad itself which may
form an ovotestis; the ducts often reflect the gonadal differentiation which
has occurred on that side of the animal. Therefore, mammals develop
according to some type of hierarchical system with the differentiation of
the gonads controlling the differentiation of the somatic tissues of the
genital ducts and genitalia. To understand the mechanisms involved in this
hierarchical system, it is first necessary to describe the changes which are
observed in sexual differentiation.

The chronology and categories of sexual differentiation

For the first third or so of the gestation period of mammals, it is not
possible to discern the sex of the embryo without making chromosome
preparations of some kind; this is therefore known as the *indifferent* period
(Austin and Short, 1972). The indifferent embryo contains a pair of
undifferentiated gonads, and two sets of genital ducts, the Wolffian
(mesonephric) ducts and the Mullerian ducts (or oviducts) which join on
to the urogenital sinus (figure 9.1). The first signs of sexual differentiation
occur in the gonads; in the male, the germ cells migrate into the central
(medullary) region of the gonad, seminiferous tubules develop and the
gonadal cortex remains undeveloped. The germ cells proliferate rapidly by
mitosis within the seminiferous tubules. Meiosis and spermatogenesis are
not observed until puberty. In the female, the germ cells migrate to the
cortex of the gonad and divide mitotically. No further ovarian differentiation occurs otherwise until after the period when tubules have been
formed in the male (so that the gonadal sex of the female is identified by
default as lack of male development during this period). Secondary sex
cords are then formed in the cortex of the gonad, and the germ cells enter
meiosis. Meiosis is arrested in the first division, resumes with the
development of the mature oocyte after puberty and is completed at
fertilization of the egg. The resting period when meiosis is not proceeding is
known as the *dictyate*.

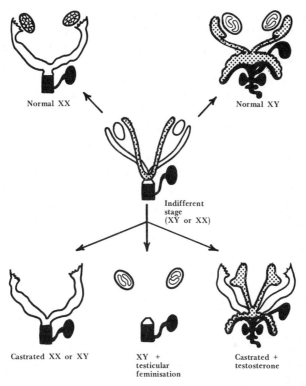

Figure 9.1 Differentiation of gonads, genital ducts and external genitalia from the indifferent stage in the mouse. Wolffian duct and derivatives are grey, Mullerian duct derivatives are outlined, and urogenital sinus derivatives are black. The paired, bean-shaped structures are the gonads, which develop as ovaries (left, above) or testes (right, above). (Modified from Ohno).

Following the initial gonadal differentiation, the genital ducts and external genitalia also change. In the male, the Mullerian duct atrophies and the Wolffian duct differentiates into the epididymis, vas deferens and seminal vesicles. The urogenital sinus differentiates into accessory organs such as prostate and preputial, and externally forms the penis. In the female the Wolffian duct atrophies whereas the Mullerian duct forms the Fallopian tubes, uterus and upper vagina; the urogenital sinus forms the lower part of the vagina and the external genitalia (figure 9.1). In the perinatal period, the female (cyclical) or male (non-cyclical) differentiation of the hypothalamus in controlling sexual cycles (by means of the hypothalamic releasing factors and their effect on pituitary gonadotrophins) is determined, although it is not expressed until puberty. This can be shown by the transplantation of ovaries into male and female animals castrated shortly after birth, and observation of the ovaries for cyclical activity after puberty (Pfeiffer, 1936). Finally secondary sexual characters (including behavioural potentialities) develop at puberty. From this simplified description, it can be seen that *gonadal sex, germ cell sex, morphological sex, hypothalamic sex* and *secondary sexual characters* differentiate in succession from the indifferent state according to the *chromosomal sex* of the individual (Austin and Short, 1972).

Genetic variation in the analysis of sexual differentiation
Studies of individuals with unusual genotypes have enabled the development of interpretations of the rules and processes governing chromosomal sex determination (Short, 1979). Firstly, there are individuals with unusual complements of the sex chromosomes, arising from errors in meiosis; secondly, mosaics or chimaeras arising naturally or artificially during development; and thirdly, alleles at specific loci which affect sexual development. These last include "sex reversal" (*Sxr*) in the mouse, resulting in male XX or XO heterozygous (*Sxr*/ +) animals; "polled" goats (resulting in intersex homozygous polled *P/P* and female heterozygous polled *P/p* XX animals); "testicular feminization" (*Tfm*), an X-linked allele described in man, the mouse, the rat and other species, which results in female genitalia combined with sterile testes in hemizygous *Tfm* XY individuals; and further X-linked loci leading to female hemizygous XY individuals both in man (with regression of the ovaries and consequent sterility) and in the wood lemming (normal fertile females). In addition, there are some species where distinctive chromosomal sex-determining mechanisms have evolved.

The questions of the nature and mode of action of these variants will be taken up in turn as they affect the successive stages of sexual differentiation. In this way a picture will be built up of the reasons for the development of these abnormalities in relation to the genetic control of the developmental pathway for sexual differentiation.

Chromosomal sex (sex determination)
The study of individuals which have an extra sex chromosome, or are deficient in a sex chromosome arising from a fault in meiosis of the parental germ cells (known as sex chromosome aneuploids), casts light on

Table 9.1 Characteristics of sex chromosome aneuploidy in man

Karyotype	Phenotype
XX	Fertile female
XY	Fertile male
XXX	Fertile female
XYY	Fertile male
XO (Turner's syndrome)	Sterile or subfertile female
XXY (Klinefelter's syndrome)	Sterile male

the role of the X- and Y-chromosomes in sexual differentiation. Thus in *Drosophila* it is found that it is the ratio of the number of X chromosomes to the number of sets of autosomes which determines phenotypic sex (Mittwoch, 1973) whilst the Y-chromosome carries loci involved in spermatogenesis (Hess, 1970). For example, XO *Drosophila* are male, but sterile. In mammals, a different set of rules operates (Austin and Short, 1972; Ohno, 1967). Individuals carrying a Y-chromosome are male whilst those without a Y-chromosome are female, but fertility also depends on possessing an appropriate number of X-chromosomes (see table 9.1). In addition to the male determining function, the Y-chromosome may also carry other loci (Stewart and colleagues, 1980). Chromosomal sex can be studied to some extent without examining the complete karyotype of the dividing cell. In the interphase nuclei, the number of Barr bodies formed from the heterochromatic X-chromosome is normally one less than the total number of X-chromosomes in accordance with the Lyon hypothesis (chapter 5). In some species, including man, the terminal portion of the Y-chromosome has a selective affinity for the fluorescent acridine dye quinacrine dihydrochloride, and can therefore be recognized in interphase nuclei using ultraviolet microscopy.

In summary, male or female development occurs in the presence or absence of the Y-chromosome respectively. For fertility, an appropriate number of X-chromosomes is also required.

Gonadal sex: roles of the somatic and germinal components of the gonad

The gonad is made up of two components. Before gonadal differentiation occurs, germ cells migrate from the posterior yolk sac region into the

genital ridge. Germ cells can be identified histologically and by their
alkaline phosphatase activity (Austin and Short, 1972). The germ cells can
be made deficient in number by either treatment with the drug busulphan
in rats (Merchant, 1975), or by the action of the W^v mutation in the mouse
(Coulombre and Russell, 1954; Mintz, 1957). Such deficiency in the germ
cells results in sterility of the affected animal, so it is generally held that the
germ cells are the precursors of the gametes. It has also been proposed that
the germ cells could play a decisive part in determining the differentiation
of a gonad as an ovary or a testis. However, the germ-cell deficient gonads
obtained in the situations described above are capable of differentiation, as
are genital ridges transplanted to the kidney capsule of host animals before
the migration of germ cells into them (Everett, 1943). Although it could be
argued that a few germ cells might still be present and sufficiently potent to
direct gonadal differentiation in each of these experimental situations, the
simpler interpretation is that the somatic component of the gonad is
directly responsible for gonadal differentiation (McCarrey and Abbott,
1979). The first expression of this is the formation of the primary sex cords
which later become the seminiferous tubules.

The critical question in gonadal development then becomes the mode of
action of the Y-chromosome in the somatic gonad, whereby its presence or
absence leads to the presence or absence of testicular differentiation.

The H-Y antigen and gonadal differentiation

Circumstantial evidence

In 1955 it was reported that there is a male-specific antigen in mice
(Eichwald and Silmser, 1955). The expression of this antigen could be
under the control of the Y-chromosome. The male-specific antigen can be
detected in mice if skin grafts are made between adult animals of an inbred
strain (which are therefore compatible for the H-2 loci). Grafts from a male
to a female are then rejected weakly, whereas grafts in the other
combinations (male to male, female to female and female to male) are
tolerated. The male donor therefore appears to possess some substance
(recognized as an antigen) not present in the host female. A similar antigen
can also be detected indirectly by using an antiserum against male cells
raised by injecting them into a compatible female. This antiserum can be
adsorbed against the tissue it is wished to type, and the degree of residual
anti-male activity present in the serum can then be determined by titration,

for example by the ability of the antiserum to kill mouse spermatozoa or lyse erythrocytes which carry the antigen (Goldberg and colleagues, 1971; Wolf and colleagues, 1980). Such sperm cytotoxic assays are technically difficult. Nevertheless it appears that all tissues tested (including spleen, liver, brain and muscle as well as skin) carry the male-specific antigen. Alternatively, in mice the H-Y antigen can also be detected in cell-mediated cytotoxic assays using mixed lymphocyte cultures from H-2 compatible animals *in vitro* (Gordon, Simpson and Samelson, 1975). It is however possible that these various techniques will recognize different male-specific antigens.

The genetic control of the male-specific antigen has been studied in man by looking at human individuals who are carrying a deletion of part of the Y-chromosome. These individuals may be male or female, depending on whether or not the person still carries the male-determining factor, located near the centromere of the Y-chromosome. It has generally been found that the male individuals are positive for male-specific antigen, and the females are negative (Koo and colleagues, 1977; Faggiano and colleagues, 1980). Hence this cytogenetic evidence indicates that the expression of the male-specific antigen is under the control of a Y-chromosome locus closely linked to the male-determining factor. XYY men also express greater levels of H-Y antigen than XY individuals (Wachtel and colleagues, 1975). The male-specific antigen is now generally known as the H-Y antigen in recognition of its control by the Y-chromosome.

The close genetic linkage of the expression of the male-determining antigen and the male-determining factor has naturally led to the specula-tion that the two loci might be identical, i.e. that the Y-chromosome might induce testicular differentiation by means of the H-Y antigen. Ohno and his colleagues (Ohno, 1979) have been particularly active in propounding the evidence for this theory. Some of the initial evidence for an association of the H-Y antigen with testis differentiation is listed in table 9.2. In animals carrying the various mutants affecting sexual differentiation, it appeared that these two factors were never dissociated (irrespective of the phenotype of the genital ducts which is discordant in testicular feminization—see below). Several groups also found correlations between the presence even of small areas of testicular tissue and the presence of H-Y antigen in mosaic individuals in man (Wolf, 1979). However, this evidence is not as convincing as it might have seemed. Correlation is not causation and the H-Y antigen could merely be a closely linked "marker" of the true male- (testis-) determining factor. Furthermore, there is no good reason to expect a complete correlation between gonadal sex and the presence or

Table 9.2   The phenotype, chromosomal sex and H-Y antigen status of various mammals. For further description of the phenotypes, see the text.

Animal	Chromosomal sex	Phenotype	H-Y antigen
Male mammals of most species	XY	Fertile male	+
Female mammals of most species	XX	Fertile female	−
Testicular feminization (mouse, man)	XY	Female genitalia but with testes	+
Sex-reversed mouse	XX	Sterile male	+
Intersex goat	XX	Sterile male or intersex	+
Pedigrees of intersex individuals (man, dog)	XX	Sterile males	+
Female wood-lemming	XY	Fertile female	−
Pure gonadal dysgenesis (man, X-linked)	XY	Sterile female	−
Female mole-vole	XO	Fertile female	−
Male mole-vole	XO	Fertile male	+

absence of the H-Y antigen in these mutants even if the theory is correct. In the proposed pathway

$$\text{Y-chromosome} \xrightarrow{(1)} \text{H-Y antigen} \xrightarrow{(2)} \text{testis}$$

a developmental block could occur in XY individuals either at stage (1) (giving the observed associations in XY females of wood lemmings or man) or at stage (2) (giving individuals who were H-Y positive, but lacking a testis). In other words, the only clear prediction from the theory was that all individuals with a testis should possess the H-Y antigen (but not *vice versa*).

A further argument has been that the H-Y antigen (or something sufficiently similar to cross-react immunologically) has been evolutionarily conserved and is always associated with the heterogametic sex (e.g. in birds, where the female possesses a non-homologous sex chromosome pair (WZ), she is H-Y positive and the male (ZZ) is H-Y negative). Evolutionary constancy is taken to imply importance (e.g. Wachtel and colleagues, 1975). But it is not clear why the H-Y antigen should evolve a female-determining function in one phylogeny and a male-determining one in the other, nor how a switch from one function to the other could occur during the evolutionary process! Until we understand why the H-Y antigen should be more than a fortuitous concomitant or marker of heterogamety, this evidence does not add to the argument for the H-Y antigen as the male-determining factor in mammalian development.

More than the circumstantial evidence given in the two preceding paragraphs is required to confirm the theory that H-Y antigen is a morphogenetic factor. Fortunately, direct experimental evidence is now forthcoming.

Experimental evidence

If organ cultures of male foetal bovine gonads at the indifferent stage are cultured with anti H-Y antiserum, this has been claimed to inhibit testicular differentiation (reviewed in Short, 1979). More recently, it has been possible to obtain a kind of H-Y antigen preparation from Daudi-Burkitt lymphoma male cell hybrids, which lack the HLA-controlled binding antigen for H-Y and therefore release it into the culture medium. This has been partially purified and a protein (molecular weight, 18 000 as a monomer) which binds to foetal bovine gonads has been characterized. If this preparation is added to bovine foetal ovary in organ culture, the ovary is transformed into a testis within five days (Ohno and colleagues, 1979). This preparation is the only known agent for inducing gonadal sex reversal in eutherian mammals. If these experiments can be repeated by others, this isolation and characterization of a cell surface antigen controlling morphogenesis is of quite exceptional importance.

Less is known about the mode of action of the H-Y antigen. Since it is present ubiquitously in the male embryo (except perhaps in the germ cells up to the late primary spermatocyte stage), the somatic tissues of the gonad must possess a receptor system for H-Y such that they respond differently from the other tissues and organs. The cytotoxic or skin graft rejection responses to the H-Y antigen are H-2 restricted in mice (i.e. they vary in strength with the particular H-2 haplotypes of the donor and responding strains of mouse), and are therefore thought to be mediated through H-2 regulated receptors on the cell surface (Hurme and colleagues, 1978). However, a separate class of receptor has recently been identified in the gonad at the indifferent stage of development, and this may be the first element in a gonad-specific receptor system (Muller and Wolf, 1979). Since the H-Y antigen is present on all tissues, it is also possible that there are other effects mediated by it yet to be discovered. In this way, it could act as a trigger to various cell-autonomous effects in different parts of the body of the male.

The regulation of expression of H-Y

We can now look at the genetic control of the expression of the H-Y antigen, and how this relates to the aetiology of the various intersex syndromes. As we have seen, the expression of the H-Y antigen is initially determined by the possession of a Y-chromosome in the normal male. This role of the Y-chromosome is confirmed by the fact that the level of the H-Y antigen is increased in XYY men. Wolf and colleagues (1980), using an assay system based on the haemolysis of male erythrocytes by anti H-Y antiserum, have found that both XO and XXp⁻ women (with a deletion of the distal part of the short arm of the X-chromosome) express some H-Y antigen. If true, this implies (i) that a certain threshold level of H-Y antigen is necessary for testis development (ii) unless all the women studied possessed Y-chromosome fragments which could not be detected, H-Y antigen is not specified by the Y-chromosome but regulated by it (i.e. the male-determining locus of the Y acts by causing the testis-inducing H-Y antigen to be synthesized from an X-linked or autosomal locus).

Cytogenetic deletion mapping implicates a locus on the X-chromosome in the region of the band known as Xp223, which Wolf and colleagues suggest would repress H-Y (in opposition to the inducing effect of the Y male-determining locus). This region of the human X-chromosome is not subject to X-inactivation (p. 85) so that both alleles in the normal XX female would repress H-Y expression. In the XO or Xp⁻ women, this repression would be partly lifted even in the absence of the Y-chromosome. In X-linked pure gonadal dysgenesis, some XY individuals show a cytological change in the same region of the X-chromosome, so that Wolf suggests that this region has become more active (perhaps being replicated several times) and represses H-Y expression despite the presence of the Y so that the individuals are H-Y negative and female. An homologous effect is postulated in the XY female wood lemming where a similar cytological alteration of the X-chromosome is observed (Herbst and colleagues, 1978).

There are several possible explanations for the various XX and XO males with no apparent Y-chromosomes. There could be undetectably small translocations of the male-determining factor of the Y-chromosome to an autosome; or, on the above model, there could be a mutation of the (autosomal) structural locus of the H-Y antigen; or a mutant allele could mimic the inducing effect of the Y-chromosome for H-Y antigen.

Very recently, it has also been reported that some women with the non X-linked form of 46, XY pure gonadal dysgenesis are H-Y positive, so that this disorder is probably heterogeneous (Wolf, 1979). This could be explained if the H-Y antigen in these cases fails to act, either because of a mutation to the morphogenetic domain of the molecule, or because of a deficiency in the gonadal receptor system (see also Nagai and colleagues, 1980, and p. 187). More questionable are reports on transsexuals, where some XY individuals originally having testes are H-Y negative, whilst some XX individuals have ovaries but are H-Y positive (Engel, Pfäfflin and Wiedeking, 1980). Whether these unusual findings have any bearing on the changes of gender identity remains to be seen, but they do not appear to be associated with the hormonal status of the individual. It is worth pointing out that this evidence, if accepted, completely undermines the circumstantial evidence given in table 9.2. Nevertheless, a model that remains compatible with the morphogenetic action of H-Y can be proposed.

If it is accepted that the structural gene for H-Y antigen is not on the Y-chromosome (Wolf, 1980; Wachtel, 1980), then there are at least three regions of the H-Y structural locus which have to be considered in the interpretation of these experiments. These are the coding sequences for the morphogenetically active site of the H-Y molecule (gonadal receptor binding site), for the antigenic site(s) of H-Y (immunologically recognized by antibodies and in cytotoxic assays), and for the regulation of H-Y expression. Depending on whether these coding regions overlap or not, and on their order within the locus, single mutations and deletions in the intersex syndromes could affect one or more of these properties simultaneously (see also Nagai and colleagues, 1980). This model is sufficiently complex to account for virtually any conceivable observation in terms of the relation between H-Y status and gonadal phenotype! It is possible that H-Y antigen as presently understood actually comprises several categories of molecule, and that the various humoral and cell-mediated immunological techniques which have been used do not all define the same category (Wachtel, 1980).

In summary, the original circumstantial evidence that H-Y is the physiological testis inducer (Table 9.2) no longer appears valid, as the equally convincing reports outlined above provide examples not only of H-Y positive individuals with ovaries, but of H-Y negative individuals with testes. Therefore, the experimental support for the hypothesis that the H-Y antigen is the mammalian testis inducer is now limited to the demonstration that a preparation which binds to foetal bovine ovaries is capable of transforming them into testes in organ culture (Ohno and colleagues, 1980). There is a clear need for the active H-Y preparation to be purified and characterized both physically and immunologically, so that the reports on genetic variants outlined above can be rigorously evaluated in relation to this substance using consistent methodological criteria. Only then will definitive models of the regulation of H-Y antigen expression, the site(s) of the structural gene(s), and their roles in testicular differentiation, be constructed.

Germ cell sex

The discussion so far has avoided the question of why individuals aneuploid for the sex chromosomes are so often sterile (e.g. in XO females and XXY males in man). On investigation, this turns out to be primarily due to a failure of the germ cells to undergo gametic differentiation, rather than the consequence of inadequate differentiation of the genital tract or deficient hormone secretion. Similarly, XX germ cells in the testis of an XX ↔ XY male mouse chimaera are non-viable (Burgoyne, 1978). This exception to the initial generalization that sexual differentiation is organized in a hierarchical system rather than autonomously has been discussed perceptively by Short (1979), who expresses it by saying that the rules for the determination of germ cell viability and sex are different from those for determination of gonadal and phenotypic sex.

In females, the lack of a second X-chromosome in XO animals leads to a decrease in the viability of the germ cells at about the time when they enter the arrested meiotic (dictyate) stage. This is the period when reactivation of the inactive X-chromosome has taken place in germ cells of normal XX females, so it is presumed that decreased levels of X-linked enzymes are responsible for the drop in viability. The degree by which germ cell viability is reduced varies in different species: in man, few if any germ cells survive, whereas in the mouse XO females are fertile. However, even here the number of germ cells is reduced and the XO females produce fewer litters of offspring than their XX sibs. The critical factor may be the absence of surviving germ cells at the time of puberty leading to an absence of follicle development and consequent endocrine deficiencies (Burgoyne, 1978). If germ cells are absent, this leads to the degeneration of the somatic component of the ovary to form a so-called "streak gonad" as a result of this inter-dependence of the germ cells (eggs) and the ovary. Hence, there is a failure of the secondary sexual characters to appear spontaneously at puberty.

Thus, the first rule of oocyte differentiation is that two X-chromosomes are necessary for fully viable oocyte development. Would the presence of a Y-chromosome interfere with oocyte development, in the way that it switches ovarian to testicular development in the somatic part of the gonad? Evidence on this point is scanty; it is difficult to obtain data on the fate of XY cells within the ovary as almost all XX ↔ XY chimaeras develop as males. However, a male mouse which derived its Y-chromosome, along with other genetic markers, from its XX ↔ XY mother has been reported (Evans and colleagues, 1977) and furthermore, XY germ cells in a foetal testis can be induced to enter meiosis prematurely (at the time this would occur in the ovary) in response to the meiotic inducer from a foetal ovary in co-culture (Byskov and Saxen, 1976). Hence, the presence or absence of a Y-chromosome is probably irrelevant to the development of the oocyte.

The phenotype of the gonad in human XY pure gonadal dysgenesis can now be explained. The primary block is in the differentiation of a testis. The gonad remains female, but the germ cells atrophy due to the lack of a second X-chromosome. In turn, the lack of germ cells leads to the formation of a streak gonad.

In the male, a wider range of evidence on the roles of the X- and Y-chromosomes in germ cell development is available. In XX ↔ XY mouse chimaeras, which are nearly all male, the use of genetic markers (e.g. for coat colours) has revealed that all the progeny are derived from the XY cell line, and none from XX cells. Cytogenetic studies reveal that the XX germ cells do not survive into the adult testis. Similar results are observed in various non-chimaeric males with two X-chromosomes (e.g. XXY in man, XX Sxr in the mouse, XX "polled" homozygotes in the goat), which are all infertile. The infertility of the majority of XXY men indicates that the extra X-chromosome, rather than the lack of the Y-chromosome from an XX germ cell, is sufficient to account for these observations (which also indicates that the second X must be active). As in females, the first rule of male gametic development is the necessity for the appropriate number of X-chromosomes (Burgoyne, 1978).

Does the Y-chromosome in mammals have a role in spermatogenesis, as it does in *Drosophila* (p. 183)? In the mouse, XX *Sxr* male mice, as we have seen, are sterile and the testes deficient in germ cells. In XO *Sxr* male mice, some germ cells survive and spermatogenesis can be seen. (This difference reflects the first rule of gametic differentiation). However, many spermatogenic cells degenerate during meiosis, relatively few spermatozoa are produced, and these have abnormal heads and are immobile. These abnormalities indicate that, in addition to the role of the Y-chromosome in the production of H-Y antigen (present in these animals) for the differentiation of the somatic component of the testis, the Y-chromosome probably plays an important role in spermatogenesis. This possible additional role is further supported by the evidence that the normal variation between inbred strains at a Y-linked locus affects the frequency of head abnormalities in mouse spermatozoa, and by the suggestion that deletion of a locus on the long arm of the Y-chromosome in man can result in a failure of spermatogenesis (Krzanowska, 1969).

A further finding related to spermatogenesis may be mentioned here. Mouse chimaeras between normal XY and *Tfm* XY eggs can be made;

from the genetic markers, the *Tfm* XY germ cells are seen to give rise to fertile spermatozoa. Hence, the germ cells do not apparently require the androgen responses mediated by the normal cytosol androgen receptor (which is absent in *Tfm* XY animals) to undergo spermatogenesis. The action of androgens on the testis which is known to be necessary for the initiation and maintenance of spermatogenesis is therefore probably mediated through the Sertoli cells which support the germ cells (Lyon and colleagues, 1975).

In summary, then, the rules for functional germ cells in the appropriate gonadal environment are:

(i)   there should be the appropriate number of X-chromosomes;
(ii) the Y-chromosome is probably necessary for normal spermatogenesis, but does not interfere with oogenesis.

The genetic control of some aspects of gametic differentiation remains unresolved. These include the determination of the germ cell number, which is probably greater in the male than the female even before the germ cells have migrated into the gonadal ridge (Beaumont and Mandl, 1963), and the genetic regulation of the factors which control the time of onset of meiosis in the developing gonad (Byskov and Saxen, 1976).

Hormonal sex and phenotypic differentiation

Historically, the recognition that sexual differentiation in mammals is organized hierarchically rather than autonomously preceded the genetic study of mouse chimaeras with which this chapter was introduced. This followed from the discovery of the effect of castration on plumage in fowls in the nineteenth century and the classical experiments on the role of the gonads in the sexual differentiation of rabbits carried out in Paris by Jost (reviewed by Jost and colleagues, 1973). Jost discovered that castration of any embryo *in utero* at the indifferent stage of development resulted in the female pattern of genital duct development, irrespective of the genetic sex of the individual. Conversely, the treatment of any embryo (whether castrated or not) with testosterone—most conveniently administered to the mother—resulted in the masculinization of the embryos (figure 9.1). Thus, the primary event in mammalian sexual differentiation was postulated to be gonadal differentiation; if male, this then imposed the male pattern of development on the rest of the body by means of testosterone secreted from the testes. In contrast to birds where males and females can differ in their response to sex steroids, the karyotype of the responding tissues is irrelevant to the result. This hypothesis has proved to be the keystone in the interpretation of mammalian sexual differentiation, and provides the explanation for the observations on natural and artificial chimaeras, where mosaic development is not normally seen.

There is however one respect in which the morphology of castrated animals treated with androgens does not resemble that of normal males; the Mullerian ducts, which persist in such animals but atrophy in the normal male. It has been postulated that this is due to the lack of another testicular hormone. Recently, glycoproteins with the ability to induce Mullerian regression have been isolated from foetal bovine testes (Picard and colleagues, 1978). In mouse $XX \leftrightarrow XY$ chimaeras, hermaphrodites may occasionally develop with an ovary on one side and a testis or ovotestis on the other; in such cases, the Mullerian duct often persists on the ovarian side, so it appears that the Mullerian inhibiting factor has a somewhat localized action. Elegant experiments by Pfeiffer (1936) showed that a further feature of sexual differentiation—the presence or absence of the hypothalamic cycles in the stimulation of gonadotrophin release which leads to oestrous cycles in the female—was determined around the neonatal stage in rats, although it is not expressed until puberty. He monitored the hypothalamic state of his experimental animals by grafting an ovary into the anterior chamber of the eye and observing it for corpus lueum formation. Androgen treatment of females, or of animals castrated at birth, inhibited the hypothalamic cycles which develop in control animals.

Since the classic experiments of Jost, many genetic studies have been carried out on the factors involved in hormonal sex differentiation. More specifically, genetic variation affecting both hormone synthesis and target response has been found. The synthesis of steroid hormones such as testosterone proceeds by a series of enzymatically controlled conversions. Genetic defects in any one of these steps can result in testosterone deficiency in varying degrees, with consequent impairment of masculinization of the foetus of correspondingly variable severity. Many such defects have been reported in man and the rat; the majority are inherited as recessive alleles, as would be expected for enzyme deficiences (see chapter 5, and reviews by Grumbach and Van Wyk, 1974 and Bullock, 1978).

Many variants affecting the responses of target organs have also been identified. In man, a syndrome (inherited as an autosomal recessive) has been reported from the Dominican Republic where XY individuals possess markedly feminized genitalia at birth with the urethra opening at the base of the very small penis. Internally, development of the Wolffian duct derivatives (such as epididymis, vas deferens and seminal vesicle) is normal. This is associated with a lack of detectable 5α-reductase activity (to convert testosterone to the metabolic 5α-dihydrotestosterone, or DHT) in biopsy cultures, so it is presumed that differentiation of the external genitalia depends specifically on DHT. Since the genital ducts and urogenital sinus appear to possess the same androgen receptor (affected by *Tfm*—see below) the reason for this is not clear. Remarkably enough, at puberty testosterone itself results in sufficient penile growth for normal sexual intercourse with ejaculation through the urethral opening at the base of the penis (Peterson and colleagues, 1977).

A second form of androgen-insensitivity is known as testicular feminization. The mutant allele (*Tfm*) is X-linked and has been reported in mouse, rat, man and several other species. A variant allele giving rise to partial androgen insensitivity has been reported in man. In the complete form, XY individuals possessing testes which secrete androgen nevertheless have a female phenotype, with a blind vagina (the lack of Mullerian derivatives being consistent with the normal production of Mullerian inhibiting factory by the testis) (figure 9.1). Injections of testosterone, or experiments in culture, have failed to produce any response to androgen, and the female phenotype can be explained on this basis (Ohno, 1971).

The *Tfm* mutation has been widely used in the investigation of the mode of hormone action (see chapter 5), and has also proved useful in investigations of the developmental and physiological roles of androgens. Two examples may be given. The mammary gland epithelium in the embryo is inhibited from the subsequent development seen in females by the action of testosterone. This could be accomplished by an action directly on the epithelium, or by an action on the underlying mesoderm followed by mesoderm-epithelium interaction of the kind characteristic of embryonic development. Culture of normal epithelium with *Tfm*/Y mesoderm, *Tfm*/Y epithelium with normal mesoderm, and controls with androgens indicated that the inhibitory action of testosterone on mammary gland is mediated through the mesoderm (Durnberger and Kratchowil, 1980).

A second example of the use of *Tfm* as an investigatory tool concerns the possible physiological role of androgens in females. Female mammals make androgens in significant amounts, especially in the adrenals, although plasma levels are lower than in males. It has been possible to produce females of the *Tfm* XO and *Tfm*/*Tfm* genotypes, the latter by means of *Tfm* XY ↔ XY chimaeras. These females have proved fertile, so that although large scale statistical comparison with normal females was not performed, it seems clear that androgens are not essential for fertility (Lyon and Glenister, 1974).

Despite all the work on *Tfm*, it is still not clear whether it represents a mutation at a structural or a regulatory locus (Bullock, 1978). There is some evidence that in the rat the normal cytosol receptor may be present, although in small amounts, suggesting the second possibility. It should also be borne in mind that the different mutant alleles in the same or different species may not be truly homologous, even though they are all X-linked. Mutations could occur in different sequences, and regulatory and structural sequences may be adjacent or intermingled (see earlier chapters).

In man, a variety of other mutant alleles have been reported as affecting the heat stability of the androgen receptor and its ability to be translocated into the nucleus, in association with some failure in masculinization. Investigation of cell cultures in these syndromes should also be useful. At the stages following interaction with chromatin, Paigen's analysis of genetic variation affecting enzyme induction has already been discussed (chapter 5).

Evolutionary considerations

Until this point, we have assumed that mammalian males are the heterogametic sex with differentiated X- and Y-chromosomes, whilst the females possess two X chromosomes. Various species have evolved where

Table 9.3 Summary of the sex chromosome constitution of species with unusual sex determining mechanism. Unusual forms of the X-chromosome are denoted by X^A and X^y (see text). F = female, M = male.

Species	Sex:	Soma		Germ cells		Gametes	
		F	M	F	M	F	M
Isoodon and *Perameles* genera		XO	XO (in adult, except somatic gonad)	XX	XY	X	X, Y
Microtus oregoni		XO	XY	XX	OY	X	O, Y
Myopus schisticolor		XX		XX		X	
		$X^A X$	XY	XX^A	XY	X, X^A	X, Y
		$X^A Y$		$X^A X^A$		X^A	
Ellobius lutescens		XO	$X^y O$	XO	$X^y O$	X, O	X^y, O

this is not true, or where the sex chromosome constitution differs in the somatic and germinal cells (table 9.3). In an earlier section, evidence was presented that female germ cells require two active X chromosomes and the males only one, whereas the mechanism of dosage compensation by X-inactivation in females gives effectively only one X-chromosome in the somatic tissues of both males and females. Alternative mechanisms for producing the same result would involve chromosome loss or duplication in somatic tissues at the time of formation of the germ cells.

In some Australian marsupials of the genera *Isoodon* and *Perameles* (the bandicoots), an X- or a Y-chromosome is lost from both female and male somatic cell lines of the pouch young a few days after birth (Hagman and Martin, 1969; Walton, 1971). The simplest process would be to lose the paternal X-chromosome in both sexes; in the female, this would then correspond functionally with the paternal X-inactivation that occurs in marsupials. A similar result is achieved in the converse manner in the creeping vole, *Microtus oregoni*, from North America. Here the female is XO at fertilization, the male XY; but the female X-chromosome is duplicated in the germ cell line, whilst the male loses his X-chromosome from the germ cell line (where it would normally constitute part of the heterochromatic inactive sex vesicle; Ohno and colleagues, 1963). The system is self-perpetuating as the oocytes produce X-bearing eggs, the spermatocytes Y- or O-bearing spermatozoa (reviewed by Short, 1972). These species therefore provide an excellent example of differential chromosome loss or duplication as an alternative to the regulation of gene expression by heterochromatin formation in the differentiation of eggs and sperm.

The XX/XY sex determining mechanism normally gives rise to a 1 : 1 sex ratio at fertilization. Alterations to the adult sex-ratio are most often seen due to the action of sex-limited lethals in one or other sex. In the wood lemming *Myopus schisticolor*, and probably other lemming species, a large excess of females is often present and fertile XY females have been described (Fredga and colleagues, 1977). These were also found to be H-Y negative (Wachtel and colleagues, 1976). It has been postulated that an X-linked mutation, associated with a cytogenetic change (Herbst and colleagues, 1978) is responsible for feminizing the XY individuals (presumably by blocking H-Y expression) and this could be homologous to the X-linked locus in pure gonadal XY dysgenesis in man described earlier. Initially, XY females of this type would probably have been subfertile because of the lack of a second X-chromosome in the germ line (like the XO mouse); the allele could still survive in the population if there were sufficient advantage to

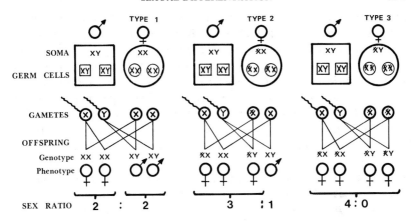

Figure 9.2 Model for the inheritance of an allele, on the X-chromosome of the wood lemming, which can suppress the effect of the male (testis)-determining locus on the Y-chromosome. (Adapted from Fredga and colleagues).

being female rather than male, allowing the evolution of a mechanism for losing the Y-chromosome by non-disjunction and duplicating the X-chromosome in the germ line (similar to that in *Microtus* XO females). There would be three types of female; normal XX, heterozygous XX^A and X^{A}Y (where X^A carries the mutant allele) which would produce a 1 : 1 sex ratio, 3 female : 1 male and all female offspring respectively (figure 9.2). The breeding data are consistent with this hypothesis. The sex ratio in the population as a whole will vary according to the relative fitnesses of the three female genotypes and of the normal XY males at the time with consequent effects on the frequency of the mutant allele.

Finally, an extraordinary situation occurs in the Asian mole-vole, *Ellobius lutescens* (Nagai and Ohno, 1977). Both males and females have XO karyotypes, but the XO males are H-Y antigen positive. This situation is most simply explained by postulating a translocation of the male-determining factor from the Y-chromosome to the X-chromosome (denoted X^y). Supposing this translocation originally occurred during embryogenesis or in spermatogenesis, the affected male would be effectively X^{y}O; when mated with a normal XX female, the offspring would have been XX^y and XO. The XX^y male would probably be sterile on account of the second X-chromosome in the germ cells, whereas the XO females might be partially fertile (as in the mouse). The only way in which the mutant X-chromosome could then survive would be if the original male mated with one or more of his daughters, giving rise to X^yO male and XO female fertile offspring. There could then be selection for increased fertility in the XO females despite the single X. Infertile XX^y males would continue to be produced to compete with fertile X^yO males for resources and female mates. Eventually, however, this genotype could become reduced to an embryonic lethal (perhaps by the loss of the X-inactivation dosage compensation mechanism in the embryo). In this way, the present system, with 50% zygotic wastage of the OO and XX^y karyotypes, would have evolved. This species therefore provides an extremely unusual example where all the individuals are seen to be descended, rather against the odds, from a single male and one or more of his daughters, with initial subfertility of the females and permanent 50% zygotic wastage. In the future, one might expect that the two X-chromosome types will diverge genetically, since they never pair with each other and cannot recombine. This is a spectacular example of the role of chance in evolution.

Sexual differentiation as a model of development

In this chapter, the process of mammalian sexual differentiation has been analysed in terms of our present knowledge of the biochemical and physiological effects of some relevant genes. This complex system provides one model for differentiation and development.

Male differentiation is triggered primarily by the action of the male (testis)-determining locus of the Y-chromosome. This switch mechanism works through a change in the cell-surface (which is recognized as the H-Y antigen), probably regulated by a complex genetic system acting in all tissues. H-Y antigen induces testicular differentiation by its action on appropriate gonadal receptors. The remainder of sexual differentiation is non-autonomous. Hormonal secretions which have either a positive effect on a variety of tissues, or a negative effect (causing atrophy of the Mullerian ducts, or, through a tissue interaction, of the mammary gland) then follow. Genetic studies reveal how the hormone acts to induce differential gene expression in a way characteristic of each tissue. This results both in the coordination of sexual differentiation throughout the body and in the amplification of relatively small differences between the tissues of the male into bigger differences (seen quantitatively in terms of the induction of new tissue-specific proteins and qualitatively in the formation of distinct tissue types; see chapter 5). X-chromosome inactivation, with chromosome loss or duplication as alternatives in some species, is one factor in the differentiation between germ cells and the somatic tissues. Genes on the Y-chromosome may be needed for spermatogenesis but not for other cell-types.

The profound differences between homologous organs in male and female mammals thus seem to depend on the presence or absence of H-Y antigen and the presence or absence of testicular hormones which results. In this analysis, many of the mechanisms that have been discussed in earlier chapters (including chromosome loss or duplication, heterochromatin formation, differential gene expression within the cell, and interactions between cells mediated by cell surface determinants, hormones and embryonic induction) can be seen to combine to produce the adult sexual phenotype.

REFERENCES

Abraham, I. and Doane, W. W. (1978) Genetic regulation of tissue-specific expression of *amylase* structural genes in *Drosophila melanogaster*. *Proc. Natl. Acad. Sci. U.S.A.* **75**, 4446–50.

Aggarwal, S. K. and King, R. C. (1969) A comparative study of the ring glands from wild type and *l(2)gl* mutant *Drosophila melanogaster*. *J. Morph.* **129**, 171–200.

Allison, A. C. (1956) Sickle cell and evolution. *Sci. Amer.* **195** (2), 87–94.

Arber, W. (1974) DNA modification and restriction. *Prog. Nucleic Acid Res. Mol. Biol.* **14**, 1–37.

Arnold, J. M. (1969) Cleavage furrow formation in a telolecithal egg (*Loligo pealii*). *J. Cell Biol.* **41**, 894–904.

Arst, H. N. (1976) Integrator gene in *Aspergillus nidulans*. *Nature* **262**, 231–34.

Artzt, K., Bennett, D. and Jacob, F. (1974) Primitive teratocarcinoma cells express a differentiation antigen specified by a gene at the T-locus in the mouse. *Proc. Natl. Acad. Sci. U.S.A.* **70**, 2988–92.

Ashburner, M. (1980) Chromosomal action of ecdysone. *Nature* **285**, 435–36.

Ashburner, M. and Richards, G. (1976) The role of ecdysone in the control of gene activity in the polytene chromosome of *Drosophila*. In *Insect Development*, ed. P. E. Lawrence, Blackwell Scientific, Oxford.

Ashton, N. W., Grimsley, N. H. and Cove, D. J. (1979) Analysis of gametophytic development in the moss, *Physcomitrella patens*, using auxin and cytokinin resistant mutants. *Planta* **144**, 427–35.

Austin, C. R. and Short, R. V. (1972) *Reproduction in Mammals*, Book 1, *Germ cells and fertilisation*; Book 2, *Embryonic and foetal development*, Cambridge University Press, Cambridge.

Avanzi, S., Maggini, F. and Innocenti, A. M. (1973) Amplification of ribosomal cistrons during the maturation of metaxylem in the root of *Allium cepa*. *Protoplasma* **76**, 197–210.

Bahr, G. F. (1970) Human chromosome fibers. *Exp. Cell. Res.* **62**, 39–49.

Beadle, G. W. and Ephrussi, B. (1937) Development of eye colors in *Drosophila*: Diffusable substances and their interrelations. *Genetics* **22**, 76–86.

Beaumont, H. M. and Mandl, A. (1963) A quantitative study of primordial germ cells in the male rat. *J. Embryol. exp. Morph.* **11**, 715–40.

Becker, H. J. (1957) Über röntgenmosaikflecken und Defecktmutationen am Auge von *Drosophila* und die Entwicklungsphysiologie des Auges. *Z. Indukt. Abstamm. Vererbungsl.* **88**, 333–73.

Beerman, W. (1952) Chromomerenkonstanz und spezifische modifikationen der Chromosomenstruktur in der Entwicklung und Organdifferenzierung von *Chironomus tentans*. *Chromosoma* **5**, 139–98.

Beerman, W. (1963) Cytological aspects of information transfer in cellular differentiation. *Am. Zool.* **3**, 23–32.

Beerman, W. (1964) Control of differentiation at the chromosomal level. *J. exp. Zool.* **157**, 49–62.

Bennett, D. (1975) The *T*-locus of the mouse. *Cell* **6**, 441–54.

Berg, R. L. (1974) A simultaneous mutability rise at the singed locus in two out of three *Drosophila melanogaster* populations studied in 1973. *Drosophila Inf. Serv.* **51**, 100–02.

Bernstein, R., Koo, G. C. and Wachtel, S. S. (1980) Abnormality of the X-chromosome in human 46, XY female siblings with dysgenetic ovaries. *Science* **207**, 768–69.

Blair, L. C., Kushner, P. J. and Herskowitz, I. (1979) Mutations of the *HMa* and *HMα* loci and their bearing on the cassette model of mating type interconversion in yeast. In *Eucaryotic Gene Regulation*, ed. R. Axel, T. Maniatis and C. F. Fox, pp. 13–26, Academic Press.

Blass, D. H. and Hunt, D. M. (1980) Pyrimidine biosynthesis in the dumpy mutants of *Drosophila melanogaster*. *Molec. gen. Genet.* **178**, 437–42.

Blass, D. H. and Hunt, D. M. (1981) Unpublished observations.

Bodmer, E. F. (1972) Evolutionary significance of the HL-A system. *Nature* **237**, 139–45.

Bonaldo, M. F., Santelli, R. Y. and Lara, F. J. S. (1979) The transcript from a DNA puff of *Rhynchosciara* and its migration to the cytoplasm. *Cell* **17**, 827–33.

Boubelik, M., Lengerova, A., Bailey, D. W. and Matousek, V. (1975) A model for genetic analysis of programmed gene expression in the development of membrane antigens. *Dev. Biol.* **47**, 206–14.

Boycott, A. E., Diver, C., Garstrang, S. L. and Turner, F. M. (1930) The inheritance of sinistrality in *Limnea peregra* (Mollusca, Pulmonata). *Phil. Trans. R. Soc. Lond. (Biol.)* **219**, 51–131.

Bradbury, E. M. (1975) Foreword: Histone nomenclature. The structure and function of chromatin. *CIBA Symp.* **28**, 1–4.

Brenner, S. (1967) The language of control. *Mem. Soc. Endocrinol.* **15**, 3–8.

Breuer, M. E. and Pavan, C. (1955) Behaviour of polytene chromosomes of *Rhynchosciara angelae* at different stages of larval development. *Chromosoma* **7**, 371–86.

Briggs, R. and Cassens, G. (1966) Accumulation in the oocyte nucleus of a gene product essential for embryonic development beyond gastrulation. *Proc. Natl. Acad. Sci. U.S.A.* **55**, 1103–09.

Brinster, R. L. (1974) The effect of cells transferred into the mouse blastocyst on subsequent development. *J. exp. Med.* **140**, 1049–56.

Brothers, A. J. (1976) Stable nuclear activation dependent on a protein synthesised during oogenesis. *Nature* **260**, 112–15.

Brothers, V. M., Tsubota, S. I., Germeraad, S. E. and Fristrom, J. W. (1978) The *rudimentary* locus of *Drosophila melanogaster*: partial purification of a carbamyl phosphate synthetase-aspartate transcarbamylase-dihydroorotase complex. *Biochem. Genet.* **6**, 321–32.

Brown, D. D. and Blackler, A. W. (1972) Gene amplification proceeds by a chromosome copy mechanism. *J. Mol. Biol.* **63**, 75–83.

Brown, D. D. and Gurdon, J. B. (1964) Absence of ribosomal RNA synthesis in the anucleate mutant of *Xenopus laevis*. *Proc. Natl. Acad. Sci. U.S.A.* **51**, 139–46.

Brown, D. D. and Sugimoto, K. (1973) The structure and evolution of ribosomal and 5S DNAs in *Xenopus laevis* and *Xenopus mulleri*. *Cold Spring Harbor Symp. Quant. Biol.* **38**, 501–05.

Brown, D. D. and Weber, C. S. (1968) Gene linkage by RNA-DNA hybridisation. II. Arrangement of the redundant gene sequences for 28S and 18S ribosomal RNA. *J. Mol. Biol.* **34**, 681–97.

Brown, S. W. and Chandra, H. S. (1973) Inactivation system of the mammalian X chromosome. *Proc. Natl. Acad. Sci. U.S.A.* **70**, 195–99.

Brutlag, D. L. (1980) Molecular arrangement and evolution of heterochromatic DNA. *Ann. Rev. Genet.* **14**, 121–44.

Bull, A. L. (1966) *Bicaudal*, a genetic factor which affects the polarity of the embryo in *Drosophila melanogaster. J. exp. Zool.* **161**, 221–42.

Bullock, L. P. (1978) Genetic variations in sexual differentiation and sex steroid action. In *Genetic Variation in Hormone Systems*, ed. J. G. M. Shire, pp. 69–87, CRC Press, Florida.

Burgoyne, P. S. (1978) The role of the sex chromosomes in mammalian germ cell differentiation. *Ann. Biol. anim. Biochim. Biophys.* **18**, 317–25.

Byskov, A. G. and Saxen, L. (1976) Induction of meiosis in fetal mouse testis *in vitro. Dev. Biol.* **52**, 193–200.

Calos, M. P. and Miller, J. H. (1980) Transposable elements. *Cell* **20**, 579–95.

Carlson, E. A. (1959) Allelism, complementation and pseudoallelism at the dumpy locus in *Drosophila melanogaster. Genetics* **44**, 347–73.

Cattanach, B. M. (1972) Evidence of non-random X chromosome activity in the mouse. *Genet. Res.* **19**, 229–40.

Cattanach, B. M. (1974) Position effect variegation in the mouse. *Genet. Res.* **23**, 291–306.

Chakrabarty, A. M. (1978) *Genetic Engineering*. CRC Press, Florida.

Chovnick, A., Gelbart, W., McCarron, H., Osmond, B., Candido, E. P. M. and Baille, D. L. (1976) Organization of the rosy locus in *Drosophila melanogaster*: evidence for a control element adjacent to the xanthine dehydrogenase structural element. *Genetics* **84**, 233–55.

Ciaranello, R. D. (1978) Genetic regulation of the catecholamine synthesizing enzymes. In *Genetic Variation in Hormone Systems*, ed. J. G. M. Shire, Vol. 2, pp. 49–61, CRC Press, Florida.

Ciaranello, R. D. and Axelrod, J. (1973) Genetically controlled alterations in the rate of degradation of phenylethanolamine N-methyltransferase. *J. Biol. Chem.* **248**, 5616–23.

Clement, A. C. (1962) Development of *Ilyanassa* following removal of the D macromere at successive cleavage stages. *J. exp. Zool.* **149**, 165–86.

Clever, U. and Karlson, P. (1960) Induktion von Puff-Veränderungen in den Speicheldrüsen-chromosomen von *Chironomus tentans* durch Ecdyson. *Exp. Cell Res.* **20**, 623–26.

Colaianne, J. J. and Bell, A. E. (1970) Sonless, a sex-ratio anomaly in *Drosophila melanogaster* resulting from a gene-cytoplasm interaction. *Genetics* **65**, 619–25.

Cole, R. K. (1967) Ametapodia, a dominant mutation in the fowl. *J. Hered.* **58**, 141–46.

Cooper, D. W., VandeBerg, J. L., Sharman, G. B. and Poole, W. E. (1971) Phosphoglycerate kinase polymorphism in kangaroos provides further evidence for paternal X inactivation. *Nature, New Biol.* **230**, 155–57.

Coleman, D. L. (1971) Linkage of genes controlling the rate of synthesis and structure of aminolevulinate dehydratase. *Science* **173**, 1245–46.

Conklin, E. G. (1931) The development of centrifuged eggs of ascidians. *J. exp. Zool.* **60**, 1–119.

Coulombre, J. L. and Russell, E. S. (1954) Analysis of the pleiotropism at the *W*-locus in the mouse: The effect of *W* and W^v substitution upon postnatal development of germ cells. *J. exp. Zool.* **126**, 277–96.

Cove, D. J., Schild, A., Ashton, N. W. and Hartmann, E. (1978) Genetics and physiological studies of the effect of light on the development of the moss *Physcomitrella patens. Photochem. and Photobiol.* **27**, 249–54.

Croce, C. M., Litwack, G. and Kaprowski, H. (1973) Human regulatory gene for inducible tyrosine aminotransferase in rat-human hybrids. *Proc. Natl. Acad. Sci. U.S.A.* **70**, 1268–72.

Crouse, H. V. (1960) The controlling element in sex chromosome behaviour in *Sciara. Genetics* **45**, 1429–30.

Crouse, H. V. and Keyl, H.-G. (1968) Extra replications in the "DNA-puffs" of *Sciara coprophila*. *Chromosoma* **25**, 357–64.

Cullis, C. A. (1977) Molecular aspects of the environmental induction of heritable changes in flax. *Heredity* **38**, 129–54.

Curtis, A. G. S. (1960) Cortical grafting in *Xenopus laevis*. *J. Embryol. exp. Morph.* **8**, 167–73.

Curtis, A. G. S. (1962) Morphogenetic interactions before gastrulation in the amphibian *Xenopus laevis*—the cortical field. *J. Embryol. exp. Morph.* **10**, 410–22.

Darlington, C. D. and Thomas, P. T. (1941) Morbid mitosis and the activity of inert chromosomes in *Sorghum*. *Proc. Roy. Soc.* **B 130**, 127–50.

Davidson, E. H. (1976) *Gene Activity in Early Development*, 2nd edn. Academic Press, New York.

Davidson, E. H. and Britten, R. J. (1973) Organization, transcription and regulation in the animal genome. *Q. Rev. Biol.* **48**, 565–613.

Davidson, E. H., Hough, B. R., Amenson, C. S. and Britten, R. J. (1973). General interspersion of repetitive with non-repetitive sequence elements in the DNA of *Xenopus*. *J. Mol. Biol.* **77**, 1–23.

Davidson, R. L., Ephrussi, B. and Yamamoto, K. (1966) Regulation of pigment synthesis in mammalian cells as studied by somatic hybridisation. *Proc. Natl. Acad. Sci. U.S.A.* **56**, 1437–40.

Deol, M. S. (1970) The origin of the acoustic ganglion and effects of the gene dominant spotting (W^v) in the mouse. *J. Embryol. exp. Morph.* **23**, 773–84.

Deol, M. S. and Whitten, W. K. (1972) Time of X chromosome inactivation in retinal melanocytes of the mouse. *Nature New Biol.* **238**, 159–60.

Dewey, M. J., Martin, D. W. Jr., Martin, G. R. and Mintz, B. (1977) Mosaic mice with teratocarcinoma-derived mutant cells deficient in hypoxanthine phosphoribosyl-transferase. *Proc. Natl. Acad. Sci. U.S.A.* **74**, 5514–68.

Dickinson, W. J. (1975) A genetic locus affecting the developmental expression of an enzyme in *Drosophila melanogaster*. *Dev. Biol.* **42**, 131–40.

Doane, W. W. (1969) *Drosophila* amylases and problems in cellular differentiation. In *Problems in Biology: RNA in Development*, ed. E. W. Hanly, pp. 75–108. University of Utah Press, Salt Lake City.

Doane, W. W. (1971) X-ray induced deficiencies of the amylase locus in *Drosophila hydei*. *Isozyme Bull.* **4**, 46–48.

Doane, W. W., Abraham, I., Kolar, M. M., Martenson, R. E. and Deibler, G. E. (1975) Purified *Drosophila* α-amylase isozymes: genetical, biochemical and molecular characterization. In *Isozymes: Genetics and Evolution*, ed. C. L. Markert, Vol. 4, pp. 585–607. Academic Press, New York.

Dobrovolskaia-Zavadskaia, N. (1927) Sur la mortification spontanée de la queue chez la souris nouveau-née et sur l'existence d'un caractère (facteur) héréditaire "non-viable". *C. R. Soc. Biol.* **97**, 114–16.

Dolnick, B. J., Berenson, R. J., Bertino, J. R., Kaufman, R. J., Nunberg, J. H. and Schimke, R. T. (1979) Correlation of dihydrofolate reductase elevation with gene amplification in a homogeneously staining chromosomal region in L5178Y cells. *J. Cell Biol.* **83**, 394–402.

Doolittle, W. F. and Sapienza, C. (1980) Selfish genes, the phenotype paradigm and genome evolution. *Nature* **284**, 601–03.

Durnberger, H. and Kratchowil, K. (1980) Specificity of tissue interaction and the origin of mesenchymal cells in the androgen response of the embryonic mammary gland. *Cell* **19**, 465–71.

Durrant, A. (1971) Induction and growth of flax genotrophs. *Heredity* **27**, 277–98.

Eichwald, E. J. and Silmser, C. R. (1955) (Skin section—no title) *Transplantation Bull.* **2**, 148–49.

Endow, S. A. and Glover, D. M. (1979) Differential replication of ribosomal gene repeats in polytene nuclei of *Drosophila*. *Cell* **17**, 597–605.

Engel, W., Pfäfflin, F. and Wiedeking, C. (1980) H-Y antigen in transsexuality, and how to explain testis differentiation in H-Y antigen negative males and ovary differentiation in H-Y positive females. *Human Genet.* **55**, 315–19.

Engelke, D. R., Ng, S.-Y., Shastry, B. S. and Roeder, R. G. (1980) Specific interaction of a purified transcription factor with an internal control region of 5S RNA genes. *Cell* **19**, 717–28.

Evans, E. P., Ford, C. E. and Lyon, M. F. (1977) Direct evidence of the capacity of the XY germ cell in the mouse to become an oocyte. *Nature* **267**, 430–31.

Evans, H. J. (1977) Facts and fancies relating to chromosome structure in man. *Adv. Human Genet.* **8**, 347–438.

Evans, G. M., Durrant, A. and Rees, H. (1966) Associated nuclear changes in the induction of flax genotrophs. *Nature* **212**, 697–99.

Everett, N. B. (1943) Observational and experimental evidences relating to the origin and differentiation of the definitive germ cells in mice. *J. exp. Zool.* **92**, 49–91.

Faggiano, M., Ferraro, M., Criscuolo, T., Sinisi, A. A. and de Capoa, A. (1980) Cytological evidence for the location of male-determining and H-Y genes on the short arm of Y chromosome. *Human Genet.* **54**, 323–26.

Falconer, D. S. (1960) *Introduction to Quantitative Genetics.* Oliver and Boyd, Edinburgh.

Fialkow, P. J. (1970) X-chromosome inactivation in the Xg locus. *Am. J. Human Genet.* **22**, 460–63.

Fielding, C. J. (1967) Developmental genetics of the mutant *grandchildless* of *Drosophila subobscura*. *J. Embryol. exp. Morph.* **17**, 375–84.

Fincham, J. R. S. and Sastry, G. R. K. (1974) Controlling elements in maize. *Ann. Rev. Genet.* **8**, 15–50.

Finnegan, D. J., Rubin, G. M., Young, M. W. and Hogness, D. S. (1977) Repeated gene families in *Drosophila melanogaster*. *Cold Spring Harbor Symp. Quant. Biol.* **42**, 1053–63.

Fischberg, M. and Blackler, A. W. (1961) How cells specialize. *Scient. Amer.* **205**, 124–40.

Ford, P. J. (1980) Polymerase III control region defined. *Nature* **287**, 109–10.

Fredga, K., Gropp, A., Winking, H. and Frank, F. (1977) A hypothesis explaining the exceptional sex ratio in the wood lemming (*Myopus schisticolor*). *Hereditas* **85**, 101–04.

Freeman, G. (1963) Lens regeneration from cornea in *Xenopus laevis*. *J. exp. Zool.* **154**, 39–65.

Gall, J. G. (1969) The genes for ribosomal RNA during oogenesis. *Genetics* **61**, Suppl. 1, 121–32.

Gallien, L., Picheral, B. and Lacroix, J.-C. (1963) Modifications de l'assortiment chromosomique chez les larves hypomorphes du Triton *Pleurodeles Waltlii* Michah obtennés par transplantation de noyaux. *C. R. Acad. Sci.* **257**, 172–73.

Gannon, F., O'Hare, K., Perrin, F., LePennec, J. P., Benoist, C., Cochet, M., Breathnach, R., Royal, A., Garapin, A., Cami, B. and Chambon, P. (1979) Organisation and sequences at the 5′ end of a cloned complete ovalbumin gene. *Nature* **278**, 428–33.

Ganschow, R. E. and Schimke, R. T. (1969) Independent genetic control of the catalytic activity and the rate of degradation of catalase in mice. *J. Biol. Chem.* **244**, 4649–58.

Garcia-Bellido, A. (1975) Genetic control of wing disc development in *Drosophila*. *Cell Patterning*, CIBA Symp. **29**, 161–78.

Garcia-Bellido, A. and Merriam, J. R. (1969) Cell lineage of the imaginal discs in *Drosophila* gyandromorphs. *J. exp. Zool.* **170**, 61–76.

Garcia-Bellido, A., Ripoll, P. and Morata, G. (1973) Developmental compartmentalization of the wing disk of *Drosophila*. *Nature New Biol.* **245**, 251–53.

Gardner, R. L. (1968) Mouse chimaeras obtained by the injection of cells into the blastocyst. *Nature* **220**, 596–97.

Gardner, R. L. (1975) Analysis of determination and differentiation in the early mammalian embryo using intra- and inter-specific chimaeras. In *Developmental Biology of*

Reproduction, 33rd Symp. Soc. Devel. Biol., ed. C. L. Markert, pp. 207–36, Academic Press.

Garen, A. and Gehring, W. (1972) Repair of the lethal developmental defect in deep orange embryos of *Drosophila* by injection of normal egg cytoplasm. *Proc. Natl. Acad. Sci. U.S.A.* **69**, 2982–85.

Garland, R. C., Satrustegui, J., Gluecksohn-Waelsch, S. and Cori, C. F. (1976) Deficiency in plasma protein synthesis caused by x-ray-induced lethal albino alleles in mouse. *Proc. Natl. Acad. Sci. U.S.A.* **73**, 3376–80.

Gartler, S. M., Chen, S., Fialkow, P. J., Giblett, E. R. and Singh, S. (1972) X-chromosome inactivation in cells from an individual heterozygous for two X-linked genes. *Nature New Biol.* **236**, 149–50.

Gateff, E. and Schneiderman, H. A. (1974) Developmental capacities of benign and malignant neoplasms of *Drosophila*. *Willhelm Roux Arch.* **176**, 23–65.

Gehring, U. and Tomkins, G. M. (1974) Characterization of a hormone receptor defect in the androgen-insensitivity mutant. *Cell* **3**, 59–64.

Gehring, W. (1973) Genetic control of determination in the *Drosophila* embryo. In *Genetic Mechanisms in Development*, ed. F. Ruddle, pp. 103–28, Academic Press.

Gehring, W. J. and Paro, R. (1980) Isolation of a hybrid plasmid with homologous sequences to a transposing element of *Drosophila melanogaster*. *Cell* **19**, 897–904.

Geyer-Duszynska, I. (1959) Experimental research on chromosome elimination in Cecidomyidae (Diptera). *J. exp. Zool.* **141**, 391–488.

Giannelli, F. and Hammerton, J. L. (1971) Non-random late replication of X chromosomes in mules and hinnies. *Nature* **232**, 315–19.

Glover, D. M. (1980) *Genetic Engineering—Clonal DNA*. Chapman and Hall, London.

Gluecksohn-Waelsch, S. (1979) Genetic control of morphogenetic and biochemical differentiation: lethal albino deletions in the mouse. *Cell* **16**, 225–37.

Goldberg, E. H., Boyse, E. A., Bennett, D., Scheid, M. and Casswell, E. A. (1971) Serological demonstration of H-Y (male) antigen on mouse sperm. *Nature* **232**, 478–80.

Goldring, E. S., Brutlag, D. and Peacock, W. J. (1975) Arrangement of highly repeated DNA of *Drosophila melanogaster*. In *The Eukaryote Chromosome*, ed. W. J. Peacock and R. D. Brock, p. 47. Australian National University Press, Canberra.

Goldstein, J. L., Marks, J. F. and Gartler, S. M. (1971) Expression of two X-linked genes in human hair follicles of double heterozygotes. *Proc. Natl. Acad. Sci. U.S.A.* **68**, 1425–27.

Gordon, R. D., Simpson, E. and Samelson, L. E. (1975) *In vitro* cell-mediated immune responses to the male-specific (H-Y) antigen in mice. *J. exp. Med.* **142**, 1108–20.

Green, M. M. (1980) Transposable elements in *Drosophila* and other Diptera. *Ann. Rev. Genet.* **14**, 109–20.

Grigliatti, T. and Suzuki, D. T. (1971) Temperature-sensitive mutations in *Drosophila melanogaster*. VIII. The homoeotic mutant ss^{a40a}. *Proc. Natl. Acad. Sci. U.S.A.* **68**, 1307–11.

Grumbach, M. M. and Van Wyk, J. J. (1974) Disorders of sex differentiation. In *Textbook of Endocrinology*, ed. R. A. Williams. Saunders, Philadelphia.

Grüneberg, H. (1969) Threshold phenomena versus cell heredity in the manifestation of sex-linked genes in mammals. *J. Embryol. exp. Morph.* **22**, 145–79.

Gurdon, J. B. (1968) Transplanted nuclei and cell differentiation. *Scient. Amer.* **219** (6), 24–35.

Gurdon, J. B. (1974) *The Control of Gene Expression in Animal Development*. Oxford University Press, London.

Gurdon, J. B. and Laskey, R. A. (1970) The transplantation of nuclei from single cultured cells into enucleate frogs' eggs. *J. Embryol. exp. Morph.* **24**, 227–48.

Hadorn, E. (1967) Dynamics of determination. In *Major Problems in Developmental Biology*, ed. M. Locke, pp. 85–104, Academic Press, New York.

Harashima, S. and Oshima, Y. (1976) Mapping of the homothallic genes, *HMα* and *HMa*, in *Saccharomyces* yeasts. *Genetics* **84**, 437–51.

Harris, H. and Watkins, J. F. (1965) Hybrid cells derived from mouse and man: artificial heterokaryons of mammalian cells from different species. *Nature* **205**, 640–46.

Harrison, P. R., Birnie, G. D., Hell, A., Humphries, S., Young, B. D. and Paul, J. (1974) Kinetic studies of gene frequency. I. Use of a DNA copy of reticulocyte 9S RNA to estimate globin gene dosage in mouse tissues. *J. Mol. Biol.* **84**, 539–54.

Hayman, D. L. and Martin, P. G. (1965) Sex chromosome mosaicism in the marsupial genera Isoodon and Perameles. *Genetics* **52**, 1201–06.

Heilig, R., Perrin, F., Gannon, F., Mandel, J. L. and Chambon, P. (1980) The ovalbumin gene family: structure of the X gene and evolution of duplicated split genes. *Cell* **20**, 625–37.

Hennen, S. (1963) Chromosomal and embryological analyses of nuclear changes occurring in embryos derived from transfers of nuclei between *Rana pipiens* and *Rana sylvatica*. *Dev. Biol.* **6**, 133–83.

Herbst, E. W., Fredga, K., Frank, F., Winking, H. and Gropp, A. (1978) Cytological identification of two X-chromosome types in the wood lemming (*Myopus schisticolor*). *Chromosoma* **69**, 185–91.

Hess, O. (1970) Genetic function correlated with unfolding of lampbrush loops by the Y chromosome in spermatocytes of *Drosophila hydei*. *Mol. gen. Genet.* **106**, 328–46.

Hess, O. and Meyer, G. F. (1968) Genetic activities of the Y-chromosome in *Drosophila* during spermatogenesis. *Adv. Genet.* **14**, 171–223.

Hewish, D. R. and Burgoyne, L. A. (1973) Chromatin substructure. The digestion of chromatin DNA at regularly spaced sites by a nuclear deoxyribonuclease. *Biochem. Biophys. Res. Commun.* **52**, 504–10.

Hicks, J. B. and Herskowitz, I. (1977) Interconversion of yeast mating types. II. Restoration of mating ability to sterile mutants in homothallic and heterothallic strains. *Genetics* **85**, 373–93.

Holliday, R. and Pugh, J. E. (1975) DNA modification mechanisms and gene activity during development. *Science* **187**, 226–32.

Hood, L. E., Wilson, J. H. and Wood, W. B. (1974) *Molecular Biology of Eukaryotic Cells. A Problems Approach*. Benjamin, California.

Hotta, Y. and Benzer, S. (1972) Mapping of behaviour in *Drosophila* mosaics. *Nature* **240**, 527–35.

Hourcade, D., Dressler, D. and Wolfson, J. (1973) The amplification of ribosomal genes involves a rolling circle intermediate. *Proc. Natl. Acad. Sci. U.S.A.* **70**, 2926–30.

Hughes, J. (1979) Opioid peptides and their relatives. *Nature* **278**, 394.

Hummel, K. P., Coleman, D. L. and Lane, P. W. (1972) Influence of genetic background on expression of mutations at the diabetes locus in the mouse. I. C57BL/KsJ and C57BL/6J strains. *Biochem. Genet.* **7**, 1–13.

Hunt, D. M. (1974) Primary defect in copper transport underlies mottled mutants in the mouse. *Nature* **249**, 852–54.

Hunt, D. M. and Johnson, D. R. (1972) An inherited deficiency in noradrenalin biosynthesis in the brindled mouse. *J. Neurochem.* **19**, 2811–19.

Hurme, M., Hetherington, C. M., Chandler, P. R. and Simpson, E. (1978) Cytotoxic T-cell responses to H-Y: mapping of the Ir genes. *J. exp. Med.* **147**, 758–67.

Illmensee, K. (1976) Nuclear and cytoplasmic transplantation in *Drosophila*. In *Insect Development*, ed. P. A. Lawrence, pp. 76–96. Blackwell, Oxford.

Illmensee, K. and Hoppe, P. C. (1981) Nuclear transplantation in *Mus musculus*: developmental potential of nuclei from pre-implantation embryos. *Cell* **23**, 9–18.

Illmensee, K. and Mahowald, A. P. (1974) Transplantation of posterior pole plasm in *Drosophila*. Induction of germ cells at the anterior pole of the egg. *Proc. Natl. Acad. Sci. U.S.A.* **71**, 1016–20.

Ilyin, Y. V., Tchurikov, N. A., Ananiev, E. V., Ryskov, A. P., Yeninkopolov, G. N., Limborska, S. A., Maleeva, N. E., Gvozdev, V. A. and Geogiev, G. P. (1978) Studies on the DNA fragments of mammals and *Drosophila* containing structural genes and adjacent sequences. *Cold Spring Harbor Symp. Quant. Biol.* **42**, 956–69.

Ising, G. and Ramel, C. (1976) The behaviour of a transposing element in *Drosophila melanogaster*. In *The Genetics and Biology of Drosophila*, ed. M. Ashburner and E. Novitski, Vol. 1b, pp. 947–54, Academic Press, New York.

Ivarie, R. D., Fan, W. J.-W. and Tomkins, G. M. (1975) Analysis of the induction and deinduction of tyrosine aminotransferase in enucleated HTC cells. *J. Cell Physiol.* **85**, 357–64.

Johnson, D. R. (1974) Hairpin-tail: a case of post-reductional gene action in the mouse egg? *Genetics* **76**, 795–805.

Jones, C. W. and Kafatos, F.C. (1980) Structure, organization and evolution of developmentally regulated chorion genes in a silkmoth. *Cell* **22**, 855–67.

Jones, K. W. and Robertson, F. W. (1970) Localisation of reiterated nucleotide sequences in *Drosophila* and mouse by *in situ* hybridisation of complementary RNA. *Chromosoma* **31**, 331–45.

Jost, A., Vigier, B., Prepin, J. and Perchellet, J. P. (1973) Studies on sex differentiation in mammals. *Recent Prog. Horm. Res.* **29**, 1–35.

Judd, B. H., Shen, M. W. and Kaufman, T. C. (1972) The anatomy and function of a segment of the X chromosome of *Drosophila melanogaster*. *Genetics* **71**, 139–56.

Judd, B. H. and Young, M. W. (1973) An examination of the one cistron: one chromomere concept. *Cold Spring Harbor Symp. Quant. Biol.* **38**, 573–79.

Kacser, H. and Burns, J. A. (1973) The control of flux. In *Rate Control of Biological Processes*, ed. D. D. Davies, pp. 65–104. Symp. Soc. Exp. Biol., Vol. 27, Cambridge University Press.

Kalthoff, K. (1971) Photoreversion of the malformation "double abdomen" in the egg of *Smittia* spec. (Diptera Chironomidae). *Dev. Biol.* **25**, 119–32.

Kalthoff, K. and Sander, K. (1968) Der Entwicklungsgang der Missbildung "Doppelabdomen" in partiell UV- bestrahlten Ei von *Smittia parthenogenetica* (Dipt. Chironomidae). *Wilhelm Roux Arch. Entwick. Org.* **161**, 129–46.

Kan, Y. W., Lee, K. Y., Furbetta, M., Augins, A. and Cao, A. (1980) Polymorphism of DNA sequence in the β-globin gene region. *N. Engl. J. Med.* **302**, 185–88.

Kandler-Singer, I. and Kalthoff, K. (1976) RNase sensitivity of an anterior morphogenetic determinant in an insect egg (*Smittia* spec., Chironomidae, Diptera). *Proc. Natl. Acad. Sci. U.S.A.* **73**, 3739–43.

Karlson, P. (1967) The effects of ecdysone on giant chromosomes, RNA metabolism and enzyme induction. *Mem. Soc. Endocrinol.* **15**, 67–76.

Kavenoff, R. and Zimm, B. H. (1973) Chromosome-sized DNA molecules from *Drosophila*. *Chromosoma* **41**, 1–27.

Kedes, L. H. (1979) Histone genes and histone messengers. *Ann. Rev. Biochem.* **48**, 837–70.

Kelly, S. J. (1975) Studies of the potency of the early cleavage blastomeres of the mouse. In *The Early Development of Mammals*, ed. M. Balls and A. E. Wild, pp. 97–105. Cambridge University Press, Edinburgh.

Kemphues, K. J., Raff, E. C. and Kaufman, T. C. (1980) Mutation in a testis-specific β-tubulin in *Drosophila*: analysis of its effects on meiosis and map location of the gene. *Cell* **21**, 445–51.

Kiss, I., Bencze, G., Fodor, A., Szabad, J. and Fristrom, J. W. (1976) Prepupal larval mosaics in *Drosophila melanogaster*. *Nature* **262**, 136–38.

Klein, J. (1975) *Biology of the Mouse Histocompatibility—2 Complex*. Springer-Verlag, New York.

Klein, J. and Hammerberg, C. (1977) The control of differentiation by the *T* complex. *Immunol. Rev.* **33**, 70–104.

Kornberg, R. D. (1974) Chromatin structure: a repeating unit of histones and DNA. *Science* **184**, 868–71.

Kornberg, R. D. and Thomas, J. D. (1974) Chromatin structure: oligomers of the histones. *Science* **184**, 865–68.

Koo, G. C., Wachtel, S. S., Krupen-Brown, K. and Mettl, L. R. (1977) Mapping the locus of the HY gene on the human Y chromosome. *Science* **198**, 940–42.

Krzanowska, H. (1969) Factor responsible for spermatozoan abnormality located on the Y-chromosome in mice. *Genet. Res.* **13**, 17–24.

Laskey, R. A. and Earnshaw, W. C. (1980) Nucleosome assembly. *Nature* **286**, 763–67.

Lewin, B. (1980) *Gene Expression.* 2nd edn, Vol. 2: *Eucaryotic Chromosomes.* Wiley, London.

Lewis, E. B. (1963) Genes and developmental pathways. *Am. Zool.* **3**, 33–56.

Lewis, E. B. and Gencarella, W. (1952) Claret and nondisjunction in *Drosophila melanogaster. Genetics* **37**, 600–01 (Abstr.)

Littlefield, J. W. (1966) The use of drug-resistant markers to study the hybridization of mouse fibroblasts. *Exp. Cell Res.* **41**, 190–96.

Lohs-Schardin, M., Sander, K., Crimer, C., Crimer, T. and Zorn, C. (1979) Localised ultraviolet laser microbeam irradiation of early *Drosophila* embryos: fate maps based on location and frequency of adult defects. *Dev. Biol.* **68**, 533–45.

Lucchesi, J. C. (1978) Gene dosage compensation and the evolution of sex chromosomes. *Science* **202**, 711–16.

Lusis, A. J. and West, J. D. (1978) X-linked and autosomal genes controlling mouse α-galactosidase expression. *Genetics* **88**, 327–42.

Lyon, M. F. (1961) Gene action in the X-chromosome of the mouse (*Mus musculus* L.). *Nature* **190**, 372–73.

Lyon, M. F. and Glenister, P. H. (1974) Evidence from *Tfm*/0 that androgen is inessential for reproduction in female mice. *Nature* **247**, 366–67.

Lyon, M. F., Glenister, P. H. and Lamoreux, M. L. (1975) Normal spermatozoa from androgen-resistant germ cells of chimaeric mice and the role of androgen in spermatogenesis. *Nature* **258**, 620–22.

Lyon, M. F., Jarvis, S. E., Sayers, I. and Johnson, D. R. (1979) Complementation reactions of a lethal mouse *t*-haplotype believed to include a deletion. *Genet. Res.* **33**, 153–61.

Lyon, M. F. and Mason, I. (1977) Information on the nature of *t*-haplotypes from the interaction of mutant haplotypes in male fertility and segregation ratio. *Genet. Res.* **29**, 255–66.

Mahowald, A. P. (1968) Polar granules of *Drosophila*. II. Ultrastructural changes during early embryogenesis. *J. exp. Zool.* **167**, 237–62.

Malacinski, G. M. and Brothers, A. J. (1974) Mutant genes in the Mexican axolotl. *Science* **184**, 1142–47.

Mandaron, P. (1971) Sur le mécanisme de l'évagination des disques imaginaux chez la *Drosophile. Dev. Biol.* **25**, 581–605.

Maniatis, T., Fritsch, E. F., Lauer, J. and Lawn, R. M. (1980) The molecular genetics of human hemoglobins. *Ann. Rev. Genet.* **14**, 145–78.

Martin, G. R. (1980) Teratocarcinoma and mammalian embryogenesis. *Science* **209**, 768–76.

Mayer, T. C. (1973) Site of gene action in steel mice. Analysis of the pigment defect by mesoderm-ectoderm recombinations. *J. exp. Zool.* **184**, 345–52.

McCarrey, J. R. and Abbott, U. K. (1979) Mechanisms of genetic sex determination, gonad sex differentiation and germ cell development in animals. *Adv. Genet.* **20**, 217–90.

McCarron, M., O'Donnell, J., Chovnick, A., Bhullar, B. S., Hewitt, J. and Candido, E. P. M. (1979) Organization of the rosy locus in *Drosophila melanogaster*: further evidence in support of a *cis*-acting control element adjacent to the xanthine dehydrogenase structural element. *Genetics* **91**, 275–93.

McCarthy, B. J. and Hoyer, B. H. (1964) Identity of DNA and diversity of messenger RNA molecules in normal mouse. *Proc. Natl. Acad. Sci. U.S.A.* **52**, 915–22.

McClintock, B. (1962) Topographical relations between elements of control systems in maize. *Carnegie Inst. Washington Yearbook* **61**, 448–61.

McCulloch, E. A., Siminovitch, L., Till, J. E., Russell, E. S. and Bernstein, S. E. (1965) The cellular basis of the genetically determined hemopoietic defect in anemic mice of genotype Sl/Sl^d. *Blood* **26**, 399–410.

McLaren, A. (1976) *Mammalian Chimaeras*. Cambridge University Press, Cambridge.

McLaren, A. and Bowman, P. (1969) Mouse chimaeras derived from fusion of embryos differing by nine genetic factors. *Nature* **224**, 238–40.

Melvold, R. W. (1971) Spontaneous somatic reversion in mice. Effects of parental genotype on stability at the *p*-locus. *Mutation Res.* **12**, 171–74.

Merchant, H. (1975) Rat gonadal and ovarian organogenesis with and without germ cells. An ultrastructural study. *Dev. Biol.* **44**, 1–21.

Metcalfe, J. A. (1971) Development and complementation of lethal mutations at the dumpy locus of *Drosophila melanogaster*. *Genet. Res.* **17**, 173–83.

Miller, J. H. and Reznikoff, W. S. (eds.) (1978) *The Operon*. Cold Spring Harbor Laboratory.

Miller, L. and Knowland, J. S. (1970) Reduction of ribosomal RNA synthesis and ribosomal RNA genes in a mutant of *Xenopus laevis* which organizes only a partial nucleolus. I. Ribosomal RNA synthesis in embryos of different nucleolar types. *J. Mol. Biol.* **53**, 321–28.

Miller, O. J., Miller, D. A. and Warburton, D. (1973) Application of new staining techniques to the study of human chromosomes. *Progr. Med. Genet.* **9**, 1–47.

Miller, O. L., Beatty, B. R., Hamkalo, A. and Thomas, C. A. (1970) Electron microscopic visualization of transcription. *Cold Spring Harbor Symp. Quant. Biol.* **35**, 505–12.

Mintz, B. (1957) Embryological development of primordial germ cells in the mouse: influence of a new mutation W^J. *J. Embryol. exp. Morph.* **5**, 396–406.

Mintz, B. (1970a) Gene expression in allophenic mice. In *Control Mechanisms in the Expression of Cellular Phenotypes*, ed. H. A. Padykula, pp. 15–42. Academic Press, New York and London.

Mintz, B. (1970b) Neoplasia and gene activity in allophenic mice. In *Genetic Concepts and Neoplasia*. Ann. Symp. on Fundamental Cancer Research, pp. 477–517. Williams and Wilkins, Baltimore.

Mintz, B. (1971) Genetic mosaicism *in vivo*: development and disease in allophenic mice. *Fed. Proc.* **30**, 935–43.

Mintz, B. and Baker, W. W. (1967) Normal mammalian muscle differentiation and gene control of isocitrate dehydrogenase synthesis. *Proc. Natl. Acad. Sci. U.S.A.* **58**, 592–98.

Mintz, B. and Illmensee, K. (1975) Normal genetically mosaic mice produced from malignant teratocarcinoma cells. *Proc. Natl. Acad. Sci. U.S.A.* **72**, 3585–89.

Mintz, B. and Sanyal, S. (1970) Clonal origin of the mouse visual retina mapped from genetically mosaic eyes. *Genetics* **64**, (Suppl.), 43–44.

Mittwoch, U. (1973) *Genetics of Sex Differentiation*. Academic Press, New York.

Monk, M. and Harper, M. I. (1978) X-chromosome activity in preimplantation mouse embryos from XX and XO mothers. *J. Embryol. exp. Morph.* **46**, 53–64.

Monk, M. and Harper, M. I. (1979) Sequential X chromosome inactivation coupled with cellular differentiation in early mouse embryos. *Nature* **281**, 311–13.

Morata, G. and Garcia-Bellido, A. (1976) Developmental analysis of some mutants of the bithorax system of *Drosophila*. *Wilhelm Roux Arch.* **179**, 125–43.

Morata, G. and Lawrence, P. A. (1975) Control of compartment development by the *engrailed* gene of *Drosophila*. *Nature* **225**, 614–17.

Moritz, K. B. and Roth, G. E. (1976) Complexity of germline and somatic DNA in *Ascaris*. *Nature* **259**, 55–57.

Muller, C. R., Migl, B., Traupe, H. and Ropers, H. H. (1980) X-linked steroid sulfatase: evidence for different gene dosage in males and females. *Human Genet.* **54**, 197–99.

Muller, U. and Wolf, U. (1979) Evidence for a gonad-specific receptor for H-Y antigen: binding of exogenous H-Y antigen to gonadal cells is independent of β_2-microglobulin. *Cell* **17**, 331–35.

Murray, K. (1978) Restriction enzymes and their uses in genetic engineering. In *Genetic Engineering*, ed. A. M. Chakrabarty, pp. 113–22. CRC Press, Florida.

Nackanishi, S., Inoue, A., Kita, T., Nakamura, M., Chang, A. C. Y., Cohen, S. H. and Numa, S. (1979) Nucleotide sequence of cloned cDNA for bovine corticotropin-β-lipotropin precursor. *Nature* **278**, 423–27.

Nagei, Y., Iwata, H., Stapleton, D. D., Smith, R. C. and Ohno, S. (1980) The testis-organising H-Y antigen of man may lose its receptor binding activity while retaining antigenic determinants. In *Testicular Development, Structure and Function*, ed. A. Steinberger and E. Steinberger, pp. 41–47. Raven Press, New York.

Nagai, Y. and Ohno, S. (1977) Testis-determining H-Y antigen in XO males of the mole-vole (*Ellobius lutescens*). *Cell* **10**, 729–32.

Namenwirth, M. (1974) The inheritance of cell differentiation during limb regeneration in the axolotl. *Dev. Biol.* **41**, 42–56.

Nöthiger, R. and Strub, S. (1972) Imaginal defects after UV-micro-beam irradiation of early cleavage stages of *Drosophila melanogaster*. *Rev. Suisse Zool.* **79**, 267–79.

Ohno, S. (1967) *Sex Chromosomes and Sex-Linked Genes*. Springer-Verlag, Berlin.

Ohno, S. (1971) Simplicity of mammalian regulatory systems inferred by single gene determination of sex phenotypes. *Nature* **234**, 134–37.

Ohno, S. (1979) *Major Sex Determining Genes*. Springer-Verlag, Berlin.

Ohno, S., Jainchill, J. and Stenius, C. (1963) The creeping vole (*Microtus oregoni*) as a gonosomic mosaic. I: The OY/XY constitution of the male. *Cytogenet.* **2**, 232–39.

Ohno, S., Nagai, Y., Ciccarese, S. and Iwata, H. (1979) Testis-organizing H-Y antigen and the primary sex-determining mechanism of mammals. *Rec. Prog. Horm. Res.* **35**, 449–70.

Okada, M., Kleinman, I. A. and Schneiderman, H. A. (1974) Restoration of fertility in sterilised *Drosophila* eggs by transplantation of polar cytoplasm. *Dev. Biol.* **37**, 43–54.

Old, R. W. and Primrose, S. B. (1980) *Principles of Gene Manipulation: an Introduction to Genetic Engineering*. Blackwell Scientific Press, Oxford.

Olins, A. L. and Olins, D. E. (1974) Spheroid chromatin units (v bodies). *Science* **183**, 330–32.

Olson, M. V., Hall, B. D., Cameron, J. R. and Davis, R. W. (1979) Cloning of the yeast tyrosine transfer RNA genes in bacteriophage lambda. *J. Mol. Biol.* **127**, 285–95.

O'Malley, B. W., Roop, D. R., Lai, E. C., Nordstrom, J. L., Catterall, J. F., Swaneck, G. E., Colbert, D. A., Tsai, M.-J., Dugaiczyk, A. and Woo, S. L. C. (1979) The ovalbumin gene: organisation, structure, transcription and regulation. *Rec. Prog. Horm. Res.*, **35**, 1–46.

O'Malley, B. W., Spelsberg, T. C., Schrader, W. T., Chytil, F. and Steggles, A. W. (1972) Mechanism of interaction of a hormone-receptor complex with the genome of a eukaryotic target cell. *Nature New Biol.* **235**, 141–44.

Oudet, P., Gross-Bellard, M. and Chambon, P. (1975) Electron microscopic and biochemical evidence that chromatin structure is a repeating unit. *Cell* **4**, 281–300.

Packman, S., Aviv, H., Ross, J. and Leder, P. (1972) A comparison of globin genes in duck reticulocytes and liver cells. *Biochem. Biophys. Res. Comm.* **49**, 813–19.

Paigen, K., Meisler, M., Felton, J. and Chapman, V. M. (1976) Genetic determination of the β-galactosidase development program in mouse liver. *Cell* **9**, 533–39.

Paigen, K., Swank, R. T., Tomino, S. and Ganschow, R. E. (1975) The molecular genetics of mammalian glucuronidase. *J. Cell Physiol.* **85**, 379–92.

Papaioannou, V. E., Gardner, R. L., McBurney, M. W., Babinet, C. and Evans, M. J. (1978) Participation of cultured teratocarcinoma cells in mouse embryogenesis. *J. Embryol. exp. Morph.* **44**, 93–104.

Pardon, J. F., Richards, B. M., Skinner, L. G. and Ockey, C. H. (1973) X-ray diffraction from isolated metaphase chromosomes. *J. Mol. Biol.* **76**, 267–70.

Pardon, J. F., Worcester, D. L., Wooley, J. C., Tatchell, K., Val Holde, K. E. and Richards, B. M. (1975) Low-angle neutron scattering from chromatin subunit particles. *Nucleic Acid Res.* **2**, 2163–76.

Pardue, M. L. and Gall, J. G. (1970) Chromosomal localization of mouse satellite DNA. *Science* **168**, 1356–58.

Perkowska, E., McGregor, H. C. and Birnstiel, M. L. (1968) Gene amplification in the oocyte nucleus of mutant and wild type *Xenopus laevis*. *Nature* **217**, 649–50.

Peterson, P. A. (1966) Phase variation of regulatory elements in maize. *Genetics* **54**, 249–66.

Peterson, R. E., Imperato-McGinley, J., Gantier, T. and Stwla, E. (1977) Male pseudo-hermaphroditism due to steroid 5α-reductase deficiency. *Am. J. Med.* **62**, 170–91.

Picard, J. Y., Fran, D. and Josso, N. (1978) Biosynthesis of labelled anti-Mullerian hormone by fetal testes: evidence for the glycoprotein nature of the hormone and for its disulfide-bonded structure. *Mol. cell. Endocrinol.* **12**, 17–30.

Pfeiffer, C. A. (1936) Sexual differences of the hypophyses and their determination by the gonads. *Am. J. Anat.* **58**, 195–224.

Port, A. E. and Hunt, D. M. (1979) A study of the copper-binding proteins in liver and kidney tissue of neonatal normal and mottled mutant mice. *Biochem. J.* **183**, 721–30.

Postlethwait, J. H. and Schneiderman, H. A. (1969) A clonal analysis of determination in *Antennapedia*, a homoeotic mutant of *Drosophila melanogaster*. *Proc. Natl. Acad. Sci. U.S.A.* **64**, 176–83.

Postlethwait, J. H. and Sneiderman, H. A. (1971) A clonal analysis of development in *Drosophila melanogaster*: morphogenesis, determination, and growth in the wild-type antenna. *Dev. Biol.* **24**, 477–519.

Quatrano, R. S. (1978) Development of cell polarity. *Ann. Rev. Plant Physiol.* **29**, 487–510.

Raff, E. C. and Raff, R. A. (1978) Tubulin and microtubules in the early development of the axolotl and other amphibia. *Am. Zool.* **18**, 237–51.

Ransom, R., Hill, W. G. and Kacser, H. (1975) Cells, clones and patches: an analysis. Cited in McLaren, A. (1976).

Rawls, J. M. and Fristrom, J. (1975) A complex genetic locus that controls the first three steps of pyrimidine biosynthesis. *Nature* **225**, 738–40.

Rendel, J. M. (1959) Canalization of the scute phenotype of *Drosophila*. *Evolution* **13**, 425–39.

Ribbert, D. (1979) Chrommomeres and puffing in experimentally induced polytene chromosomes of *Calliphora erythrocephala*. *Chromosoma* **74**, 269–98.

Richards, G. (1978) Sequential gene activation by ecdysone in polytene chromosome of *Drosophila melanogaster*. VI. Inhibition by juvenile hormones. *Dev. Biol.* **66**, 32–42.

Ritossa, F. (1973) Crossing-over between X and Y chromosomes during ribosomal DNA magnification in *Drosophila melanogaster*. *Proc. Natl. Acad. Sci. U.S.A.* **70**, 1950–54.

Royal, A., Garapini, A., Cami, B., Pessin, F., Mandel, J. L., LeMeur, M., Bregegegre, F., Gannon, F., LePennec, J. P., Chambon, P. and Kourilsky, P. (1979) The ovalbumin gene region: common features in the organisation of three genes expressed in chick oviduct under hormonal control. *Nature* **279**, 125–32.

Russell, L. B. (1961) Genetics of mammalian sex chromosomes. *Science* **133**, 1795–803.

Russell, E. S. (1970) Abnormalities of erythropoiesis associated with mutant genes in mice. In *Regulation of Hematopoiesis*, ed. A. S. Gordon, pp. 649–75. Appleton, New York.

Sakano, H., Rogers, J. H., Hüppi, K., Brack, C., Traunecker, A., Maki, R., Wall, R. and Tonegawa, S. (1979) Domains and the hinge region of an immunoglobulin heavy chain domain are encoded in separate *DNA* segments. *Nature* **277**, 627–33.

Sakonju, S., Bogenhagen, D. F. and Brown, D. D. (1980) A control region in the center of the 5S RNA gene directs specific initiation of transcription. I. The 5′ border of the region. *Cell* **19**, 13–25.

Sánchez, F., Natzle, J. E., Cleveland, D. W., Kirschner, M. W. and McCarthy, B. J. (1980) A dispersed multigene family encoding tubulin in *Drosophila melanogaster*. *Cell* **22**, 845–54.

Scharf, S. R. and Gerhart, J. C. (1980) Determination of the dorsal-ventral axis in eggs of *Xenopus laevis*: complete rescue of UV-irradiated eggs by oblique orientation before first cleavage. *Dev. Biol.* **79**, 180–98.

Scott, R. W. and Frankel, F. R. (1980) Enrichment of estradiol-receptor complexes in a transcriptionally active fraction of chromatin from MCF-7 cells. *Proc. Natl. Acad. Sci. U.S.A.* **77**, 1291–95.

Shannon, M. P. (1972) Characterization of the female-sterile mutant almondex of *Drosophila melanogaster*. *Genetica* **43**, 244–56.

Short, R. V. (1972) Germ cell sex. In *Edinburgh Symposium on the Genetics of Spermatozoa*, ed. R. A. Beatty and S. Gluecksohn-Waelsch, pp. 325–45. (Dept. of Animal Genetics, University of Edinburgh, Scotland).

Short, R. V. (1979) Sex determination and differentiation. *Brit. med. Bull.* **35**, 121–27.

Shire, J. G. M. (1976) The forms, uses and significance of genetic variation in endocrine systems. *Biol. Rev.* **51**, 105–41.

Shire, J. G. M. and Hambly, E. A. (1972) The adrenal glands of mice with hereditary pituitary dwarfism. *Acta Pathol Microbiol. Scan.* **A81**, 225–28.

Shur, B. D. (1977) Cell-surface glycosyltransferases in gastrulating chick embryos. I. Temporally and spatially specific patterns of four endogenous glycosyltransferase activities. *Dev. Biol.* **58**, 23–39.

Shur, B. D. and Bennett, D. (1979) A specific defect in galactosyltransferase regulation on sperm bearing mutant alleles of the T/t locus. *Dev. Biol.* **71**, 243–59.

Silver, L. M., Artzt, K. and Bennett, D. (1979) A major testicular cell protein specified by a mouse T/t-complex gene. *Cell* **17**, 275–84.

Silvers, W. K. (1956) Pigment cells: Occurrence in hair follicles. *J. Morph.* **99**, 41–55.

Solari, A. J. (1971) Experimental changes in the width of chromatin fibers from chicken erythrocytes. *Exp. Cell Res.* **67**, 161–70.

Sonneborn, T. (1970) Determination, development and inheritance of structures in cell cortex. *Symp. Int. Soc. Cell Biol.*, vol. 9, pp. 1–14. Academic Press, New York.

Spear, B. B. and Gall, J. G. (1973) Independent control of ribosomal gene replication in polytene chromosomes of *Drosophila melanogaster*. *Proc. Natl. Acad. Sci. U.S.A.* **70**, 1359–63.

Spradling, A. C. and Mahowald, A. P. (1980) Amplification of genes for chorion proteins during oogenesis in *Drosophila melanogaster*. *Proc. Natl. Acad. Sci. U.S.A.* **77**, 1096–1100.

Stern, C. (1968) *Genetic Mosaics and Other Essays*. Harvard University Press.

Stern, C. and Tokunaga, C. (1967) Nonautonomy in differentiation of pattern-determining genes in *Drosophila*. I. The sex-comb of eyeless-dominant. *Proc. Natl. Acad. Sci. U.S.A.* **57**, 658–64.

Sturtevant, A. H. (1929) The claret mutant type of *Drosophila simulans*: a study of chromosomal elimination and cell lineage. *J. Wiss. Zool.* **135**, 323–56.

Stewart, A. D., Manning, A. and Batty, J. (1980) Effects of Y-chromosome variants on the male behaviour of the mouse, *Mus musculus*. *Genet. Res.* **35**, 261–68.

Strobel, E., Dunsmuir, P. and Rubin, G. M. (1979) Polymorphism in the chromosomal locations of elements of the 412, *copia* and 297 dispersed repeated gene families in *Drosophila*. *Cell* **17**, 429–39.

Sullivan, D., Palacios, R., Stavnezer, J., Taylor, J. M., Faras, A. J., Diely, M. L., Summers, N. M., Bishop, J. M. and Schimke, R. T. (1973) Synthesis of a deoxyribonucleic acid sequence complementary to ovalbumin messenger ribonucleic acid and quantification of ovalbumin genes. *J. Biol. Chem.* **248**, 7530–39.

Suzuki, Y., Gage, L. P. and Brown, D. D. (1972) The genes for silk fibroin in *Bombyx mori*. *J. Mol. Biol.* **70**, 637–49.

Tarkowski, A. K. (1961) Mouse chimaeras developed from fused eggs. *Nature* **190**, 857–60.

Tartof, K. D. (1971) Increasing the multiplicity of ribosomal RNA genes in *Drosophila melanogaster*. *Science* **171**, 294–97.

Tobin, S. L., Zulauf, E., Sánchez, F., Craig, E. A. and McCarthy, B. J. (1980) Multiple actin-related sequences in the *Drosophila melanogaster* genome. *Cell* **19**, 121–31.

Tonegawa, S., Maxam, A. M., Tizard, R., Bernard, O. and Gilbert, W. (1978) Sequence of a mouse germ-line gene for a variable region of an immunoglobulin light chain. *Proc. Natl. Acad. Sci. U.S.A.* **75**, 1485–89.

Varley, J. M., Macgregor, H. C. and Erba, H. P. (1980) Satellite DNA is transcribed on lampbrush chromosomes. *Nature* **283**, 686–88.

Vasil, I. K., Ahuja, M. R. and Vasil, V. (1979) Plant tissue culture in genetic and plant breeding. *Adv. Genet.* **20**, 127–215.

Wachtel, S. S. (1980) Where is the H-Y structural gene? *Cell* **22**, 3–4.

Wachtel, S. S., Koo, G. C., Breg, R., Elias, S., Boyse, E. A. and Miller, O. J. (1975) Expression of H-Y antigen in human males with two Y-chromosomes. *New Engl. J. Med.* **293**, 1070–72.

Wachtel, S. S., Koo, G. C., Ohno, S., Gropp, A., Dev, V. G., Tantravalsi, R., Miller, D. A. and Miller, O. J. (1976) H-Y antigen and the origin of XY female wood lemmings. *Nature* **264**, 638–39.

Wachtel, S. S., Ohno, S., Koo, G. C. and Boyse, E. A. (1975) Possible role for H-Y antigen in the primary determination of sex. *Nature* **257**, 235–36.

Waddington, C. H. (1953) Genetic assimilation of an acquired character. *Evolution* **7**, 118–26.

Wakasugi, N. and Morita, M. (1977) Studies on the development of F_1 embryos from interstrain crosses involving DDK mice. *J. Embryol. exp. Morph.* **38**, 211–16.

Walton, S. M. (1971) Sex chromosome mosaicism in pouch young of marsupials, *Perameles* and *Isoodon. Cytogenet.* **10**, 115–20.

Waring, G. L., Allis, C. D. and Mahowald, A. P. (1978) Isolation of polar granules and the identification of polar granule-specific protein. *Dev. Biol.* **66**, 197–206.

Watson, G. and Paigen, K. (1978) Segregation of genetic determinants for murine glucuronidase synthesis and loss in CXB recombinant-inbred strains. *Biochem. Genet.* **16**, 897–903.

Weintraub, H. and Groudine, M. (1976) Chromosomal subunits in active genes have an altered conformation. *Science* **193**, 848–56.

West, J. D. (1975) A theoretical approach to the relation between patch size and clone size in chimaeric tissue. *J. theor. Biol.* **50**, 153–60.

West, J. D. (1976a) Distortion of patches of retinal degeneration in chimaeric mice. *J. Embryol. exp. Morph.* **36**, 145–49.

West, J. D. (1976b) Clonal development of the retinal epithelium of mouse chimaeras and X-inactivation mosaics. *J. Embryol. exp. Morph.* **35**, 445–61.

West, J. D. (1976c) Patches in the livers of chimaeric mice. *J. Embryol. exp. Morph.* **36**, 151–61.

West, J. D., Frels, W. I. and Chapman, V. M. (1977) Preferential expression of the maternally derived X chromosome in the mouse yolk sac. *Cell* **12**, 873–82.

Whittaker, J. R., Ortolani, G. and Farinella-Ferruzza, N. (1977) Autonomy of acetylcholinesterase differentiation in muscle-lineage cells of ascidian embryos. *Dev. Biol.* **55**, 196–200.

Wilson, E. B. (1925) *The Cell in Development and heredity*, 3rd edn. Macmillan Inc., New York.

Winking, H. (1981) Possible viability of mice with maternally inherited T^{hp}. *Hereditas* **94**, 19.

Winter, C. E., de Bianchi, A. G., Terra, W. R. and Lava, F. J. S. (1977) Relationships between newly synthesised proteins and DNA puff patterns in salivary glands of *Rhynchosciara americana. Chromosoma* **61**, 193–206.

Wolf, U. (1978) Zum Mechanismus der Gonadendifferenzierung. *Bull. Schweiz. Akad. Med. Wiss.* **34**, 357–68.

Wolf, U. (1979) XY gonadal dysgenesis and the H-Y antigen. *Human Genet.* **47**, 269–77.

Wolf, U., Fraccaro, M., Mayerova, A., Hecht, T., Maraschio, P. and Hameister, H. (1980a) A gene controlling H-Y antigen on the X chromosome. Tentative assignment by deletion mapping to Xp223. *Human Genet.* **54**, 149–54.

Wolf, U., Fraccaro, M., Mayerova, A., Hecht, T., Zuffardi, O. and Hameister, H. (1980b) Turner syndrome patients are H-Y positive. *Human Genet.* **54**, 315–18.

Worcel, A. and Benyajati, C. (1977) Higher order coiling of chromatin. *Cell* **12**, 83–100.

Wu, C., Wong, Y.-C. and Elgin, S. C. R. (1979) The chromatin structure of specific genes: II. Disruption of chromatin structure during gene activity. *Cell* **16**, 807–14.

Yamada, T. (1977) Control mechanisms in cell-type conversion in newt lens regeneration. In *Monographs in Developmental Biology*, ed. A. Wolsky, Vol. 13. Karger, Basel.

Yen, P. H., Sodja, A., Cohen, M., Conrad, S. E., Wu, M., Davidson, N. and Ilgen, C. (1977) Sequence arrangement of tRNA genes on a fragment of *Drosophila melanogaster* DNA cloned in *E. coli. Cell* **11**, 763–77.

Züst, B. and Dixon, K. E. (1977) Events in the germ cell lineage after entry of the primordial germ cells into the genital ridges in normal and UV-irradiated *Xenopus laevis. J. Embryol. exp. Morph.* **41**, 33–46.

Zwilling, E. (1956a) Genetic mechanism in limb development. *Cold Spring Harbor Symp. quant. Biol.* **1**, 349–54.

Zwilling, E. (1956b) Interaction between limb bud ectoderm and mesoderm in the chick embryo. IV. Experiments with a wingless mutant. *J. exp. Zool.* **132**, 241–53.

Zwilling, E. and Hansborough, L. (1956c) Interaction between limb bud ectoderm and mesoderm in the chick embryo. III. Experiments with polydactylous limbs. *J. exp. Zool.* **132**, 219–39.

Index

215

R2.1.7.3.